D0387375

CAT FLAPS
AND
MOUSETRAPS

CAT FLAPS
AND
MOUSETRAPS

BY

HARRY OLIVER

metro

Published by Metro Publishing,
an imprint of John Blake Publishing Ltd,
3 Bramber Court, 2 Bramber Road,
London W14 9PB, England

www.blake.co.uk

First published in hardback in 2007

ISBN: 978-1-84454-474-5

All rights reserved. No part of this publication may be reproduced,
stored in a retrieval system, or in any form or by any means, without the
prior permission in writing of the publisher, nor be otherwise circulated in
any form of binding or cover other than that in which it is published
and without a similar condition including this condition being imposed
on the subsequent publisher.

British Library Cataloguing-in-Publication Data:

A catalogue record for this book is available from the British Library.

Design by www.envydesign.co.uk

Printed in Great Britain by Mackays of Chatham Ltd, Chatham, Kent

1 3 5 7 9 10 8 6 4 2

© Text copyright Harry Oliver 2007

Papers used by John Blake Publishing are natural, recyclable products
made from wood grown in sustainable forests. The manufacturing processes
conform to the environmental regulations of the country of origin.

Every attempt has been made to contact the relevant copyright-holders,
but some were unobtainable. We would be grateful if the appropriate
people could contact us.

For Joey

ALSO BY HARRY OLIVER

March Hares and Monkeys' Uncles
The bestselling book on the origins of the words
and phrases we use every day

Black Cats and April Fools
Origins of the old wives' tales and superstitions
in our daily lives

Available in all good bookshops, priced £9.99.
To order a copy directly visit blake.co.uk

CONTENTS

ACKNOWLEDGEMENTS

My first thanks are to my fiancée, Joanna Kennedy, whose patient, dedicated help with researching this book was another exhibition of her selfless kindness. Your ideas, Excel spread sheets and nice coffees were just magic. Also to Ann Wilson, for your encouragement and for giving us a wonderful place to live (and work). We'll be seeing more of each other now, I hope!

Also to Liz Kennedy, for your house, your biscuits and for being great company while I was writing in Farndale at Dave Earnshaw's cottage. What a wonderful place to be productive, thanks Dave. And to Bernard Kennedy, for allowing me to take over his office, which no doubt deprived him of valuable online-flight-booking time. Bigger thanks for *that* pint of Tetley, Mr Kennedy.

Adam Parfitt still agrees to take the time out of his busy writer's life to edit what I come up with, and I can't think of a nicer man for the job. Thanks Ad.

Finally to Mike Mosedale, for the ever-wonderful illustrations, and to Graeme Andrew, of Envy Design, for laying out the book so brilliantly.

INTRODUCTION

Take a look around you and pick out any manmade object. Now consider the word 'manmade', how often we use it without thinking, and remind your self that anything manmade is exactly that – *made by man* (or woman, of course). In doing this, you have found the essence of this book's focus – the amazing things man has come up with, and the why and how of the objects we so often take for granted.

We all know the phrase 'necessity is the mother of invention', and indeed it often is. But there are other forces that bring about great things. Sometimes it is the desire for money, for fame; sometimes it is the need to help, and sometimes huge creative leaps come about by pure chance, by accident. All of these factors, and more, crop up time and time again in this book, as do stories of the blood sweat and tears of the inventors.

Many inventions seem indispensable, without them our lives would be more difficult by a long stretch. Others are luxuries that have enriched the way we live and communicate. Others we may not even be aware of. But each has its own unique history, and many of their 'life stories' make for a great anecdote.

Just as when I wrote *March Hares and Monkeys' Uncles*, which dealt with the origins of everyday words and phrases, and *Black Cats and April Fools*, which did the same for old wives' tales and superstitions, my principle aim in writing this book was to inform and at the same time entertain, and to try to avoid being tedious. When researching the innumerable inventions surrounding us, I was by turns amused, enthralled, surprised and sometimes overwhelmed by the sheer amount of information on the subject – 'How on earth can I hope to cover everything?' was a question that popped up more than once!

The only answer to this crushing poser was to be selective – of course I would try and include the common objects a reader may (quite rightly) expect to find an explanation for, but I would not be ruled by this aim. Naturally objects such as the television, the telephone, and the wheel are included, but there are also many quirkier entries, for I found that it was often the seemingly less significant inventions that had the most diverting tales to tell. So, I have ended up with a book that covers a wide mix of objects, most importantly those whose development is interesting and memorable. To anyone disappointed by any omissions, I can only apologise!

One might think who invented what should be a fairly cut and dried matter, and often it is. Frequently though, the precise origin of something is shrouded in myth, controversy, or is so old as to be ambiguous to say the least. Often, it comes down to arguments over who got there first, or indeed which inventor got to the patent office before another! In these cases I have tried to arrive at a story that conforms to the general consensus on the matter. If I need to be set straight on anything do email my publisher: words@blake.co.uk.

The things that surround us can sometimes seem mundane and a little too familiar. But writing this book, and learning of the human endeavour and drama that lies beneath even the tawdriest of objects, has brought so many of them to life for me. In the process it has enriched my experience of the modern world. I only hope reading it will do the same for you! Have fun!

CHAPTER ONE:
LEISURE AND FUN

Frisbee

LEISURE AND FUN

Barbie Doll

Whether or not Barbie and Ken represent realistic or appropriate role models for today's increasingly overweight, inactive young boys and girls is debatable, but the answer to whether or not kids like them is found in their enduring popularity – over a billion sales to date. The dolls were named after the son (Ken) and daughter (Barbara) of the co-founders of the toy and game company Mattel, Elliot and Ruth Handler. They teamed up with Harold Mattson in order to expand on their picture-framing company in 1945 and start a sideline in toys. Observing that her children preferred playing with adult dolls, and that the only adult dolls around were one-dimensional cardboard figures, Ruth came up with the idea of a more fully realised adult doll. Though rejected at first by the board, Mattel eventually came around to the idea when she pointed out the success of a similar idea in Germany. In 1959, Barbie dolls went on sale, the first one sporting a black and white zebra-striped swimsuit and signature topknot ponytail, and they were made available as either blondes or brunettes. She sold over 300,000 in the first year of production, and two years later Ken showed up. First marketed as a 'teenage fashion model', Barbie's appearance has changed many times, most notably in 1971 when the doll's eyes were changed to look forwards instead of sideways. Barbie has 38 pets – including cats and dogs, horses, a panda, a lion cub and a zebra – and she holds a pilot's licence to boot!

Chess

There is much debate about the precise origins of chess. Most people agree, however, that it evolved from the Ancient Indian war game Chaturanga, a sixth-century Sanskrit word meaning 'four parts'. Indeed, that the Portuguese, Arabic, Persian, Greek and Spanish terms for chess all derive from Chaturanga makes a strong case that India is where chess was born.

Not that the original game much resembles that of today. Originally conceived for four players, and involving dice, it was based on the Ancient Indian army, with a king, a counsellor, elephants, horses, chariots and infantry. Chess underwent many stages of transformation to become the game of skill that we recognise. By the seventh century, the game had spread to Persia, and reached Western Europe via the Moorish invaders in the eighth century. From then the modern game became established.

Cigarettes

The first observations of smoking were made in 1492 by Rodrigo de Jerez while on Christopher Columbus's expedition to the Americas. To him, natives appeared to be drinking smoke from something shaped like a 'musket formed of paper'. Rodrigo indulged in a puff or two of tobacco that had been wrapped in palm. However, when he returned to Spain he was imprisoned for having scared people with the smoke that poured from his nose and mouth! He served a seven-year sentence, and when he got out of jail smoking pipes and cigars had become common in Spain.

Smoking arrived in Britain in the 1560s thanks to Sir John Hawkins and his cousin Sir Francis Drake, who famously introduced pipe-smoking to Sir Walter Raleigh in 1585. But cigarettes akin to those people smoke today did not come about until much later – invented in 1832 at the Battle of Acre during the Turkish–Egyptian war by an Egyptian soldier. The story goes that, to increase his firing rate, the gunner had taken to rolling gunpowder in paper tubes. When his tobacco pipe broke, he used the same paper to roll tobacco with, thus creating the first cigarette.

The first manufactured cigarettes were produced in France by the manufacturer Française des Tabacs in 1843. All cigarettes were handmade and were therefore very limited luxury goods. Production of handmade cigarettes did not begin in Britain until 1856, and the first brand of cigarette made and sold in Britain was Sweet Threes, launched in 1859. The world's first mass-production factory started up in Cuba where steam-driven machines were used to create the product that would become the world's number-one consumer killer.

Crossword

The idea for the crossword came to Englishman Arthur Wynne in 1913. Employed in the 'tricks and jokes department' of the *New York World*, he was one day trying to come up with yet another diverting puzzle for his readers when he was struck by the memory of a game he used to play with his granddad called Magic Square or Double Acrostic. The game formed the basis for the first crossword, or 'Word-Cross' as he called it. There were 32 clues and the words were separated by black spaces, and it was published in the *New York World* on 21 December 1913.

The crossword went through several rather dull stages of innovation, but an important development occurred when the list of clues became two lists – horizontal and vertical – the brainchild of CW Shepherd, who first saw his crossword format appear in the *Sunday Express* on 2 November 1924. The craze for crosswords in Britain was a phenomenon – so widespread was their popularity that the British Optical Association voiced worry that they could strain readers' eyes and bring on headaches. These concerns were largely ignored, and the craze continued unabated.

The first cryptic crossword was compiled for the *Saturday Westminster* in 1925 by 'Torquemada', or Edward Mathers, who took for his pseudonym the name of a spirited Spanish Inquisitor. The cryptic variety delighted English puzzlers, but not their American counterparts. The most cryptic crossword ever devised was the work of Sir Max Beerbohm, who fulfilled his fantasy of creating a puzzle 'with clues signifying nothing – nothing whatsoever'!

During the war, several crosswords in the *Daily Telegraph* came under the scrutiny of Allied security officers, as words that were secret codenames used in Operation Overlord kept occurring. The crossword compiler Leonard Dawe was arrested and investigated, but it was concluded that it was all a coincidence down to troops stationed in the area using the words in passing while in the presence of the innocent Dawe's children, who then repeated the words to their father. It has since been claimed that Dawe himself picked up the words by eavesdropping on soldiers around army camps, but the matter remains a bit of a puzzle, and there are no clues as to what the correct answer is.

Fireworks

'Remember remember, the fifth of November' goes the English rhyme, referring to the infamous Guy Fawkes and his thwarted attempt to blow up Parliament in 1605. To celebrate his failure, Guy Fawkes night sees hundreds of firework displays across the UK. But fireworks were around long before we started using them.

Originating in China, the story goes that gunpowder came about by accident around 2,000 years ago when a Chinese cook chanced to mix the then common kitchen materials charcoal, sulphur and saltpetre. It burned well and, when packed into a bamboo tube, exploded. A thousand years later, gunpowder was used by Chinese monk Li Tian, who was living in the Hunan province in the vicinity of Liuyang City, to create the firecracker – today the Liuyang region produces a vast amount of the world's fireworks. The loud explosions were (and still are) thought to be effective in warding off evil spirits. His invention is annually celebrated in China on 18 April, and firecrackers still feature hugely in Chinese New Year celebrations to embrace a new year with an absence of any nasty spirits.

Some sources say the Crusaders brought gunpowder back from their travels, but explorer Marco Polo is largely given credit for introducing it to Europe in the 13th century. Initially it was used in weapons – rockets, canons and guns in particular – but it was the Italians who first produced fireworks. They get a mention in Shakespeare, and Elizabeth I was so fond of them that a new post – Fire Master of England – was created! Today, fireworks continue to delight humans, and terrify dogs and cats, the world over.

Frisbee

It is perhaps surprising to think that the origin of the Frisbee is linked to the world of pies, especially because one would not normally associate pie-eating types with the sorts who like to run after a bit of plastic spinning in the air. However, without the pie we would not have the Frisbee, as it was Connecticut baker William Russell Frisbie whose pie tins were embossed with his name. His pies were hugely popular with New England college students, and they soon discovered that the empty pie tins flew gracefully through the air. Thus, a new game was born on American campuses, although debate still continues among Frisbee obsessives as to which college housed 'he who was the first to fling'. Some at Yale College have the audacity to suggest that an 1820 undergraduate called Elihu Frisbee threw a chapel collection tray across campus and, in doing so, invented the Frisbee. But most agree it was the pie-throwers.

The first commercial, pie-tin-like disc was produced in 1948 by Walter Morrison, who cited a popcorn can lid that he tossed around one Thanksgiving Day as his inspiration to create. He eventually produced the Pluto Platter, and sold the rights to Richard Knerr, of Wham-O toys. On hearing the Frisbie baking story, Knerr cleverly capitalised on it, changing the Pluto Platter into the Frisbee (note the altered spelling) and shrewdly marketing it as a new sport in 1964. Unsurprisingly, sales soared …

Hula Hoops

Standing around twirling a plastic ring around your waist isn't so popular with kids these days, perhaps due to the proliferation of other, perhaps more interesting pursuits like texting and violent video games. However, back in the 1950s the activity was a positive craze, with children everywhere desperate to get their hands (or waists) on American company Wham-O's groovy plastic toy – the idea for the hoops was brought to them from Australia, where children used bamboo hoops, and by the end of the decade they had already sold 100 million of them! Children of other nations were less fortunate – they were banned in Japan because of all that indecent hip-wiggling, and the Russian government weren't so keen either. What none of these eager hoop-fiends knew was that their trendy game was nothing new – hoops have been played with ever since ancient times, though then they were made of grapevines and tough grass rather than plastic. In

11

14th-century England, 'hooping' was popular, but deemed bad for backs and hearts and soon banned. Hooping became 'hula hooping' when 18th-century British sailors observed the similarity between the Hawaiian 'hula' dance and the gyrating movements required to keep a hoop revolving round a waist!

Jigsaw

It was the combination of John Spilsbury's map-making and engraving skills that brought about the invention of the jigsaw. During the 1760s, Spilsbury started making some maps aimed at children to develop their knowledge of geography. He thought that a smart way for children to memorise the position of England's counties in relation to each other would be to employ his engraving skills. He mounted the maps on to wood to reinforce them, and then he neatly scored round the counties to separate them for the purposes of an exercise where the shapes were removed and mixed in order for England to be pieced back together again. Spilsbury made jigsaws commercially available in the late 1960s. Jigsaw puzzles were originally made using a jigsaw which started out as a hand saw but the modern version is a power tool with an electric motor.

Jukebox

For the record, it is worth noting that music boxes and players were in use before jukeboxes. Located in fairgrounds and amusement arcades, they were coin-operated and played one tune per 'pay'. But the arrival of the jukebox truly blew them out of the water. The first jukebox only used an electric Edison phonograph which was connected to four listening tubes with a 'nickel in the slot' device attached to each one. The jukebox was installed by Louis Glass in 1889 in San Francisco at the Palais Royal Saloon. In 1905, a 24-track jukebox arrived on the scene, the precursor of jukeboxes as we think of them. Each recording was on a cylinder. Disc-playing jukeboxes followed shortly, with a 1906 model playing 24 10in discs. Various stages of development led to the 45rpm vinyl record jukebox in the 1930s, manufactured by the Seeburg Corporation, and this became the industry standard until the arrival of CD jukeboxes in the 1980's. The 'juke' in 'jukebox' comes from the African-American slang word 'jook', which means 'dance'.

Kite

Kite-flying was not invented as a summer activity to go alongside a picnic in the park. The motivation was a little more practical than that. The kite was used by the Chinese military to carry messages over long distances – the information was conveyed by the colour of the material used. In the latter part of the 18th century, after being introduced into Europe, kites were used to measure wind speed, and then by the Wright brothers for carrying out preliminary investigations into flight theory before launching the first powered aircraft.

The fascination with holding on to a piece of canvas blowing around in the sky hasn't died yet. Proof of this is the popularity of kite-surfing and stunt kites. The advent of massive 'power kites' has enabled a person (often a single man with few friends) to become airborne for a few seconds at the risk of a severe ankle sprain on crashing back to the ground. This is always amusing for smug couples strolling in the park.

Lego

Lego has been a part of children's toy boxes for the last 50 years. Endless varieties of structure can be erected with the brightly coloured plastic building blocks, and it is a hardwearing toy that lasts for generations.

Lego was the mastermind of a skilled craftsman, yet the Danish carpenter who invented it was producing something quite different to plastic toys: wooden ladders! It was after the Wall Street Crash of 1929, when the Depression ensued, that the impoverished Ole Kirk Christiansen was inspired to come up with possible sidelines to make some extra cash. He was a creative, family-oriented man, and he set about making children's toys from offcuts of wood left over from making ladders. He began ambitiously, coming up with some rather fancy toys. Kids loved his creations, but were more impressed with the wooden bricks he had begun to construct. They found them a more interactive choice of toy, and they could build whatever they wished, which saved Ole a task! Christiansen named his new company Lego. Lego is a hybrid of *leg godt*, which means 'play well'.

Lego progressed from making building blocks out of wood to making them out of plastic. Each block was designed with a hollow underside so that one could slot neatly on top of the next. Rather in the way that Lego pieces get passed from one generation to the next, the Lego business did the same. Ole's son Godtfred Kirk Christiansen not only took up where his father left off, but also made a major improvement to the design. To the hollow underside of each unit he added cylindrical bits of plastic that tightly connected with studs on top of the brick and held each structure together more firmly, yet still allowed them to be pulled apart for the building of another structure.

The Christiansens patented the idea and the portfolio was extended to include larger bricks for younger children. They also added plastic Lego figures for children to add life to their creations, and so it went on.

Monopoly

For years no household was complete without a Monopoly board, and no family game of Monopoly was complete without the spectacular fallings-out that tend to happen when relatives congregate to see who can be the most ruthlessly successful property magnate. Thirty years before Charles Darrow became famous for inventing the version of the Monopoly board game we play today, Lizzie Magie was the woman who created the original idea in 1904. Magie was firmly opposed to capitalism and wanted the game to highlight the potential misery caused by buying and selling habits, and to point a finger at the perpetual wealth generated by real-estate moguls. She called it the Landlord's Game and was granted a patent that year. Charles

Darrow picked up where she left off and made his own version, reversing the name of the game to Capitalism, for which he was also granted a patent in 1935. The game came out of a holiday in Atlantic City, a place he and his wife frequented, when Darrow devised a game that incorporated names of streets there. He modelled the playing figures that everybody is fond of on the charms hanging from his wife's bracelet. He was undeterred when at first his creation was rejected by the big games companies, so he went back to his workshop and worked away at producing 5000 copies of the game. It finally made the shops in the USA in 1935 and by 1936 a UK version with London streets was made available. Today, Monopoly is sold in 80 countries and produced in 26 languages. To date over 200 million games have been sold worldwide.

Rubik's Cube

The Rubik's Cube, also known as the magical cube, was the inspiration of Hungarian architect Ernö Rubik. Rubik's hobby was sculpting, but he had a particularly keen interest in geometry and multi-dimensional forms. During one of his many experiments, he stacked a number of little blocks together into a larger cube and placed different brightly coloured stickers on each cube. He became fascinated with the smaller blocks and the many possible positions they could occupy within the larger cube. When he tried to get the cubes to return to their original positions within the larger cube, he realised he had a real puzzle on his hands. In 1975 his application to patent the toy was accepted and the 'Magical Cube', as it became known in Hungary, went on sale in 1977. The toy was later judged formally and went on to win the German Toy of the Year award in 1980. By 1982 over 100 million cubes had been sold. Different-sized versions of the same cube began to be manufactured and in 2006 international Rubik's Cube champion Frank Morris solved the even more challenging 7x7x7 version of the cube, invented by Greek (and possibly geek) Panagiotis Verdes, in just under 6½ minutes.

Scrabble

It was during the 1930s that out-of-work architect Alfred Butts started to think about the immense popularity of certain games. He realised that games of chance were particularly popular, so thought about incorporating a random element into a game. He also wanted to devise a game that would focus on vocabulary skills, as he was an avid reader of the *New York Times* and knew how popular the crossword had become (see page 6). He decided to come up with a game that was half luck, half skill. His first idea was Lexico, similar to the Scrabble that we know today, but without a board upon which to arrange the tiles. His idea was turned down by the most influential games companies, so he thought about making it more substantial and tightening up the rules a bit. He decided the game would be better played upon a board so that contestants could sit around and have a central focus. He also scored each letter by tallying the frequency with which it occurred from one page to the next in the *New York Times*. He called it Criss Cross (which referred both to the criss-crossing squares on the board and the word 'crossword'). He then teamed up with James Brunot, who had the legal know-how and marketing skills to get the game on the market in 1946. Brunot renamed it Scrabble, and from then on all Butts had to do was collect the royalties. In the early 1950s Brunot sold it on to Selchow & Righter, a games company better equipped to deal with mass-production, and they sent its popularity into orbit. Scrabble is now sold in over 120 countries.

Slinky

Richard James was a struggling nautical engineer working in a shipyard in Philadelphia when he stumbled upon his abstract idea for a toy. Walking past a table on the ship, he happened to see an abandoned spring component from an engine roll from the table. He was amazed to see the spring travel onwards across the deck in incremental rolling jumps, and immediately had the idea for a toy of some sort. He went home and set about copying the helical design in order to work out the best dimensions for it, and in which material it would work the most smoothly. It ended up being about the size of an orange in diameter and seemed to work most effectively when made from metal. His wife Betty catchily named the toy the Slinky after seeing the word, which means 'sleek and graceful', in the dictionary. In 1948, the pair produced and boxed up hundreds of the neat little toys to take to a presentation in a department store where they demonstrated the Slinky's potential to a large audience. Slinky fans snapped up the Jameses' stocks so quickly that they rushed home and set about opening their own company with plans to build a factory to cope with rising production levels.

Swimming Pool

Although there is evidence to suggest crudely built swimming pools were around as early as 3000 BC, it was in Rome that Emperor Gaius Maecenas built the first heated pool. It was fuelled by a furnace-based central-heating system. The Romans used swimming pools for training the military and preparing athletes for competitive events. They were also used for relaxation regimes. It was the Ólympics that brought the swimming pool craze to Britain in 1896. The following year several pools were constructed with diving boards and facilities, and swimming has continued to be popular ever since.

Roller Skates

In 1760, the first attempt at roller skating didn't go well. Joseph Merlin, a Belgian violin-maker, showed up to the London masquerade party of the famous Mrs Cornelly. Steaming into the Soho ballroom on skates he had designed himself, he played the violin with aplomb as he soared across the room. Seconds later he found himself unable to stop or turn and, according to a report from the time, 'impelled himself against a large mirror valued at over £500, smashed it to atoms, broke his instrument, and wounded himself severely'. Unsurprisingly, Joseph's new leisure pursuit didn't catch on fast. In fact, it didn't catch on until over 60 years later when another skating pioneer, Robert John Tyers, showed off his 'Volitos' in 1823. In some way resembling modern inline skates, or rollerblades, Tyers's skates had five wheels in a straight line to be attached to each foot. There were no nasty accidents, and he garnered a few followers, but it was not until 1863, when New Yorker James L Plimpton patented his four-wheeled roller skates with a rubber cushion above each wheel, allowing the skater to turn a smooth curve by simply leaning to one side, that the roller-skating craze swept America and then the world.

Teddy Bear

Morris Michtom, a toy-shop owner in Brooklyn, made the first teddy bear. His creation was inspired by an incident involving President Theodore Roosevelt. In 1902 Roosevelt, a fan of hunting, joined a group on a shoot. The day was a bit of a flop and, towards the end of it, Roosevelt was ushered in the direction of a badly injured bear cub that his fellow hunters thought he would enjoy being able to shoot outright. The president was appalled by the suggestion and said he couldn't possibly kill the poor lame cub. After the trip the story was soon leaked to the papers, and numerous features – including a strip by cartoonist Clifford Berryman – were plastered across the newspapers. In the same year Morris Michtom was the first to jump on this idea commercially when he made a soft-toy version of the injured bear. Michtom named his creation 'Teddy's Bear' and even politely wrote to the president seeking permission to sell it under his name. Permission was granted and the comforting, much-loved figure sold in its thousands.

Yo-yo

It is said that the word yo-yo originates from a dialect spoken in the Philippines during the 19th century, and that it means 'come back' – it is rumoured to have started out there as a weapon that was sometimes made more lethal with attachments of blades and studs. They were then used for catching prey or as a means of defence. Evidence of people enjoying yo-yo-like toys recreationally can be seen in Greek sketches dating back to 500 BC. Though popular in the centuries that followed, it wasn't until the 1930s that the market really exploded as a result of some clever marketing by one Pedro Flores, who opened a factory that focused solely on yo-yo production. Various other entrepreneurs got involved in the action and such was the huge popularity of the yo-yo by the 1930s that over 300,000 units a day were being produced by just three US factories.

The yo-yo developed over the years, improvements chiefly aiming to minimise friction so that string-winding speed and ease could be optimised. The development that really excited yo-yoers around the world was the addition of the ball bearing to the axle, which helped the yo-yo to spin on the end of the string for amazingly long periods of time, thus allowing increasingly fancy tricks to be performed. Yo-yo geeks from around the globe meet annually in Florida to show off their spinning skills.

CHAPTER TWO:
AROUND THE HOME

'Dammit, the burglar alarm's gone off on it's own again.'

Burglar Alarm

AROUND THE HOME

Aerosol Can

The use of the aerosol can arose out of a pressing need and, as unpleasant as body odour may be, it was not the need for a spray-on antiperspirant deodorant! Rather, it was a matter of life and death. Based in the South Pacific during the Second World War, a large number of American troops were dropping dead from malaria-infected mosquitoes and other nasty airborne creatures, and the Department of Agriculture was desperate to combat this. In 1943 two Americans, Lyle Goodhue and William Sullivan, came up with the 'bug bomb', an aerosol insecticide. They based their invention on a model for a pressurised container that, years earlier in 1926, had been invented by Norwegian Erik Rotheim. Rotheim's invention was refillable and had a valve, a carbon dioxide propellant system and a nozzle. Thanks to him, the American troops were able to get on with their business without the itchiness and, moreover, death caused by mozzie bites.

After the war, improved aerosols were marketed for a multitude of purposes (including deodorant), and it was not until the 1970s, when the environmentally damaging effects of chlorofluorocarbons became apparent, that aerosol design had to be reconsidered. Nowadays, aerosol cans use carbon dioxide, nitrous oxide and hydrocarbons.

Aluminium Foil

Where would last night's leftovers and tomorrow's sandwiches be without good old aluminium foil, or tin foil as most of us mistakenly call it (due to the fact that it was first made out of a combination of tin and lead). Originating from the Reynolds Metal Company in Louisville, Kentucky, it was produced by Richard S Reynolds, who took advantage of falling aluminium prices in the 1920s and began producing foil for tobacco producers and confectionary makers. By 1947, the company had devised its most well-known creation, Reynolds

Wrap Aluminum Foil, designed for the domestic market. It transformed the way American housewives stored their food, and the brand still sells very well across the world today.

Battery

The Italian physicist Alessandro Volta is the man behind the invention of the modern battery. In 1800, it came about because of a highly charged argument with his colleague Luigi Galvani, a medical doctor. During physiological experimentation, Galvani observed that two different metals, in contact with a dead frog's dissected leg, happened to produce a little charge of electricity. Galvani's colleague Volta wished to prove that the charge of electricity arose due to a continuous circuit formed between the metal and the frog's leg. Volta set about showing this by substituting brine-soaked cardboard for the frog flesh. Discs made from copper and zinc were layered alternately and connected to the cardboard in the salt water, which was there to improve the conducting power. Volta connected wires to the metals, through which a continuous current was successfully passed. His 'voltaic pile' inspired the modern electrochemical battery, and his name, Volta, formed the basis of the name for the unit of electric potential difference, the volt.

Burglar Alarm

It seems that, at the time of the electric burglar alarm's invention, ordinary American folk in the 1850s were more fearful of electricity than they were of being burgled, for, although Bostonian Edwin Holmes came up with a crafty gadget that rang a bell when a door or window was opened, nobody wanted to use it. And anyway, there wasn't much crime in Boston at the time! Perhaps you can steal a guess as to why he found more commercial success on moving to New York …

Candle

Early forms of the lamp were around in Africa as early as 65,000 BC, tending to consist of bowls made from rock into which animal-fat-soaked materials were placed and burned. Following this, floating wicks in oil were used. Solid candles did not emerge until roughly 3000 BC, when beeswax was shaped into a column with a wick running through the middle. Candle design was revolutionised in the early 1800s by French chemist Michel Chevreul, who prepared stearin (a glyceryl ester that comes from beef fat) from stearic acid and used it as the basis for his invention, stearin candles, which burned less smoke, were brighter and remained firmer than the waxy variety.

Cat Flap

Most of us think of Isaac Newton as the man who first defined the laws of gravity when an apple fell from a tree on to his head, but the great man has more than this to his name – he also invented the first known cat flap! Isaac was very fond of his cat but, much to his annoyance, it kept nudging the door to his laboratory open and allowing unwanted sunshine in while he was conducting light experiments. Not having the heart to banish the little creature from the lab, he made a hole in the door and covered it in felt, so the cat could come and go as it pleased. The story goes that, when his cat had kittens, Sir Isaac made smaller flaps for them in the same door, overlooking the fact they could have used the bigger flap anyway. An absent-minded bit of DIY while on a break from inventing the third law of motion perhaps!

Central Heating

In the days of the Roman Empire, the equivalent memory of hugging a radiator on returning from a winter walk might have been that of spreading oneself across a tiled floor to warm up. Under-floor heating, known as a hypocaust, is the earliest documented central heating system. The hypocaust worked by diffusing heated air from a furnace, distributing it into empty spaces between pillars that supported floor tiles. The air was also released through holes in the walls. Though under-floor heating is still very popular in Japan and Korea, from the 16th century it began to be replaced by the seemingly less sensible radiator, which circulates heated water rather than air and only heats things locally to it.

Chair

Chairs are nothing new, and may well have been around well before the first examples we know of, which date to around 2600 BC. These ornate seats were discovered in the tombs of Ancient Egyptian kings, and were there in the first place so that the kings had somewhere nice and comfy to relax in the afterlife. As you might imagine, they had lovely padded seats – no expense spared!

But, despite their long-standing (or -sitting) place in history, chairs have not always been the common items we see them as today. Until relatively recently (well, the 16th century), chairs were almost exclusively used for special people in special situations – they denoted stature and power. Far more common for ordinary use until then was the stool or bench.

Dishwasher

Josephine Cochrane was a lady who didn't like washing up. In fact, she hated it, but would not trust her servants to treat her china with care. To get around this she spent ten years developing a machine to do the job for her. In 1886 she filed a patent for her invention, but was unable to perfect it until her husband passed away. He was a bit of a meany and had not been prepared to help her out financially, presumably because – dishwashing machine or no dishwashing machine – he never had to scrub those pots himself and had little to gain should such a device come about. But, once he was out of the way, there was no stopping Mrs Cochrane – now a moneyed widow – from producing machines for domestic and commercial use. Home dishwashers relied on a handle which, on being turned, drove pistons that pumped water and soap around the machine several times before the dishes were left to drain. Dishwashing machines in hotels were powered by steam.

In 1889 the manufacturing rights were sold to the Crescent Washing Machine Company, an American outfit. So, next time you load up your beloved machine after dinner, you've got one woman to thank. Mrs Cochrane, you were lazy when it came to doing the dishes, yet admirably driven and determined when it came to thinking up an alternative. Bless you for the trouble you've saved us!

Dry-cleaning

The ability to clean clothes without water was discovered by accident by French tailor Jolly Bennin. One day in 1849 he had the (mis)fortune to knock over a spirit lamp on to one of his wife's freshly laundered tablecloths. As he tried to cover up the spillage before his wife noticed, he observed that the spirit-soaked sections were actually cleaner than the rest of it. This led him to conduct experiments that lead to a whole new branch of his business, which he called *Nettoyage à sec*. Clothes were unstitched, then placed in turpentine oil, brushed, dipped again and dried before being sewn up again. A revolution in cleaning techniques ensued. Soon, a more effective spirit was developed that allowed clothes to be cleaned without being unstitched, and in 1866 the first dry-cleaning service, Pullars of Perth, began doing business by post in Britain.

Electric Iron

Invented in 1882 by American Henry W Seely, the first electric iron used two carbon rods that were housed in the iron's base. Electricity passed over the gap between the rods and heated the footplate of the iron. A year later, he came up with the cordless iron, which was designed to sit in a stand until hot enough to use. This was much safer and hassle-free for the user, but the problem was that nobody was able to use Henry's marvellous invention at first, because electricity was not yet available! It wasn't long, though, before power stations came along and changed all that, and by 1891 electric irons were being sold by the British companies GEC and Crompton.

The 1920s saw the emergence of the first steam irons in the USA, produced by Eldec, a dry-cleaning company, but it was not until 1936 that technology caught up enough to produce the first electric steam irons for domestic use. Since then, they've been improved upon, and improved upon again. Today, with more efficiency than ever, electric irons continue to make easier the miserable, soul-destroying task of removing creases from our clothes ...

Flushing Toilet

The popular belief that Thomas Crapper was the inventor of the flushing toilet, and that his name was the origin of the word 'crap', is a false one – a load of crap, even. He *was* a plumber, and *did* own a toilet factory, but he had nothing to do with the development of the lavatory.

In fact, flushing toilets of sorts were in existence long before Mr Crapper would ever have needed the loo. As early as 2000 BC in Crete, the Minoans developed a toilet that relied on gravity – stream water flowed into stone cisterns and the water was released when a lever was pulled. Similarly, the Romans relied on gravity to flush water through their magnificent waste drainage systems.

But the flushing toilet more along the lines of those we are used to was invented by Sir John Harington, godson of Elizabeth I, in 1597. Amusingly, due to the indecency of the subject of toileting, she would not grant Sir John a patent for his invention, the Ajax (a pun on 'a jakes', then a slang term for toilet), but was more than happy to have one installed in her Richmond residence! This hypocrisy is perhaps unsurprising for, after all, she was notoriously obsessed with cleanliness, famously taking a bath once a month 'whether she needed it or not'! Whether Britain needed flushing toilets or not, commercial production only began nearly two centuries later in 1778, after Englishman Alexander Cumming patented his S-bend design, and improvements were made by Joseph Bramah, whose company sold toilets for over a century. Bramah toilets were made of metal and brass, which were gradually superseded by ceramic models.

Food Processor

Although electric mixers were already being used by commercial bakers, it was not until 1919 that the first food mixer for domestic use came about. Named the Kitchen Aid, it was based on a brilliant design by American Robert Johnson. Its 'planetary action' – a revolving bowl, and beaters that revolve in the opposite direction to it, is not a feature of modern mixers, but influenced the development of food processors, which did much more than simply mix. The first blender was produced in 1922 by American Stephen J Poplawski, who dispensed with beaters suspended above the bowl, instead placing a spinning blade at the bottom. In 1950, Ken Wood developed the first – somewhat cumbersome – food processor, which set new standards in quick chopping and mixing. But the man credited with the invention of the modern, compact and lightweight multipurpose food processor is French engineer Paul Verdun. Working for a catering company as a salesman, it is perhaps unsurprising that his observations regarding the amount of time people spent slicing and dicing in the kitchen could be better used for other more interesting pursuits. He oversaw the development of a machine that could chop, whisk, blend, slice (you name it!) and the Magimix went on sale in 1971. It has since been imitated the world over.

Fork

Ancient Middle Easterners used forks with two prongs, but there is no evidence to suggest that the fork arrived in Europe any earlier than the 11th century. The first mention of the fork is found in a story from this time. While in Italy, the wife of a Byzantine Emperor shamed herself by insisting on using the little fork she'd brought along on her travels because she didn't care for eating with her hands. The church very much looked down on the fork and did everything they could to prohibit its use. What we now consider an invaluable utensil was seen as distasteful and sacrilegious because using one did away with the use of fingers for eating

– fingers were, of course, gifts from God and seen as nature's forks! People therefore tried to limit fork usage by only shaking off sauces or dressings from bits of meat with them. Once this had been accomplished, they would take the flesh from the fork and eat it with their hands. But they couldn't win: according to the church, this fork behaviour wasn't acceptable either.

Two-pronged forks were gradually superseded by four-pronged forks for reasons of efficiency, and over the next few hundred years using a fork at dinner gained popularity in Europe. By the 17th century, having once been the cause of such moral outrage, it had come full circle – the fork had become a hit with the upper classes and for quite a long time it was only the well-moneyed who could afford to fork out for a fork!

Light Bulb

Depending which side of the Atlantic you're on, there's a chance you could make someone incandescent with rage (or at least raise the odd defensive eyebrow) when naming the inventor of the light bulb, for an American and an Englishmen are both cited. In America, credit is given to Thomas Edison, and in England Joseph Swan wins the day. The fact of the matter is that both played a part in its development.

As early as 1801, Englishman Sir Humphrey Davy showed us that when electricity was passed through a filament it glowed enough to produce light. The problem was that, in contact with air, the filaments did not last long enough because they burned up very quickly. The crucial task became finding a way to keep a filament away from air, and many inventors scratched their heads for years, producing a number of bulbs that didn't fit the bill. In 1879, Joseph Swan cracked the problem (but luckily not the bulb) and showed the world his magnificent invention, which used a carbonised cotton thread as a filament.

Across the pond, however, Edison had also been looking into the same problem, and not long after Swan he too came up with a bulb that used carbonised paper filaments. Being the clever chap he was, he patented the bulb there and then. In 1880 the first light bulbs became available to the public, and Edison had beaten Swan to it – Swan's bulbs didn't come out until a few months later. When Edison went for a British patent too, Swan took him to court, but eventually Swan and Edison saw there was no need to argue and teamed up to form the Edison and Swan

United Electric Company. Employing glass blowers from Germany (the Brits were not skilled enough!), the two lit up the world together.

Matchstick

Burned bone remains found in Africa suggest that the usefulness of fire was known of over a million years ago. In the past, fires were started using laborious methods such as rubbing sticks together. It wasn't until the late 17th century that a way of producing fire spontaneously and quickly was discovered. Robert Boyle, an English physicist, produced an early version of the matches we use today when he dipped sticks into sulphur and phosphorous and rubbed them together to produce a flame. These were very crude and unsafe matches, carrying a high risk of explosion. This made them impractical because they could not be carried around. In 1827 English chemist John Walker accidentally started a small fire when working with the compounds potassium chlorate and antimony sulphide, which he had mixed together with a stick. Absent-mindedly, he scraped the mixing stick across an abrasive stone surface: the friction produced caused a rise in temperature sufficient enough to ignite the mixture of the compounds and they burst into flames. Walker's match was the first strikeable match and became known as the 'friction match'. For some reason he didn't apply for the patent, which enabled other chemists to delve in and develop the match. Several others experimented with the efficiency and safety of the friction match, and in 1855 Johan Lundstrom from Sweden came up with a safe design that prevented spontaneous ignition. The tip of Lundstrom's match did not possess all the chemicals required for producing a flame; instead they were split between the match and the striking strip on the matchbox.

Microwave Oven

Whether for heating up last night's Chinese takeaway, zapping a ready meal or steaming the broccoli, no modern kitchen is complete without a microwave. Surprisingly, though, the first commercial microwave was available in America back in 1955 for a cool $1,300. More surprising still, its invention was a bit of an accident.

In 1946, American Percy Spencer was busying himself working on a piece of aviation radar equipment called a magnetron when he discovered that the chocolate bar he'd been looking forward to had melted in his pocket. Percy supposed that the microwave radiation emitted from the magnetron may have been to blame for this and, more than a little intrigued (and perhaps in need of another snack now his chocolate had melted), he placed some maize in front of the machine. Sure enough, he soon had a bag's worth of popcorn! Convinced he was on to something, he bored a hole in a kettle and aimed a microwave beam through it. Hey presto, he had created the first microwave oven! Delighted, he demonstrated his discovery to a colleague by placing an egg into the kettle, and within seconds the egg exploded over both of them. Treading on eggshells at first, he persuaded the company Raytheon to develop a less primitive version, and in 1946 a patent was filed. In 1947, the first microwave for use in catering stood at a massive six feet tall and cost $5,000. Technology has moved on a little since then …

Milk Carton

The invention of the Tetra Pak is not the most electrifying of stories, but it very significantly revolutionised the way we pack and preserve food. In 1943 Ruben Rausing, of the Swedish package manufacturer Akerlund & Rausing, began researching ways to pack liquid cheaply and efficiently. Rausing wanted to make a carton out of strong paper and he tried several geometric variations on a flat-pack shape to work out which would stand the most sturdily while also being easy to reproduce. (He also worked out the best method of reproduction, constructing them straight from a continuous roll of paper and immediately filling them.) The tetrahedron, a figure with four triangular faces, worked the best and a plastic clip served well as an effective seal. The idea was patented in 1944, the name being a combination of the words 'tetrahedron' and 'packaging'. However, a suitable coating to make the paper waterproof was needed before production could begin. Once this was sorted, the first milk cartons were in circulation in 1951 and, before long, the Tetra Pak was being used for all manner of liquids.

Mousetrap

For American William Hooker and Englishman James Henry Atkinson, the question of how to catch a mouse was a burning one. While the people of two nations held their brooms aloft and endured their pantries being raided by the elusive little buggers, these fellows sat down in their respective countries, thinking caps firmly on. Hooker was the first to patent his 'Out O Sight' trap in 1894, and Atkinson followed in 1899 with his 'improved' sprung mousetrap, suggesting he had copied Hooker to some degree. The difference between the two is that Hooker's model merely required the mouse to walk across the trap, whereas Atkinson's device used the bait on a hook we are all familiar with. He called it the 'Little Nipper', which is manufactured to this day in Wales by Proctor Brothers.

Teflon and the Non-stick Pan

Roy Plunkett, researcher at polymer company DuPont, discovered Teflon in 1938. Plunkett was busy fiddling with tetrafluorethylene gas in a large tube, trying to assess its suitability as a coolant in fridges. Having used dry ice to store the gas in small cylinders to prepare it for chlorination, on opening the cylinders, he was most surprised to discover that the gas had turned to solid – all that remained was a white powder. It turned out that the iron-coated interior had acted as a catalyst in a reaction that formed the polymer polytetrafluorethylene (PTFE). He tested PTFE, finding it to be a very large molecule with a melting point of more than 250°C. Other potentially advantageous properties discovered were that it was dense, hardy and resistant to heat, extreme weather and corrosive chemicals – nothing less than a wonder product! The new material merited immediate patenting by DuPont in 1941, and Teflon was registered for trademark in 1944. Many uses were found for Teflon, but perhaps the most famous product associated with it is the non-stick pan. There is a popularly held belief that non-stick pans were a by-product of the US space programme, but this is a myth. While the space programme made extensive use of Teflon, it was in the late 1950s that French engineer Marc Grégoire had the brainwave of exploiting Teflon's high melting point and low-friction properties in the kitchen. He attached PTFE to the base of kitchen pans and changed the way we fry forever.

Refrigerator

Based on the principle of cooling through evaporation, a University of Glasgow-based chemist named William Cullen achieved freezing conditions in 1775 by evaporating ether (changing it from a liquid to a gas), thus lowering the surrounding temperature, because when a liquid evaporates, the surface it evaporates from cools because the rising molecules take heat with them. Quite a time passed before American Jacob Perkins put this discovery to use and took it a step further. He added sulphuric ether into the mix, and evaporated this and ether, then compressing the gas to return it to its liquid state for re-use. He applied for and received a patent for mechanical refrigeration in 1834, although he did not go on to develop a commercial product. From the 1850s onwards, the food and drink industry embraced refrigeration for the purposes of production, but domestic refrigerators did not go on sale until 1911 when General Electric produced a rather cumbersome, motor-powered machine. In the mid-1930s, the introduction of chlorofluorocarbons enabled much more convenient and efficient units to be produced, and they became ubiquitous in households throughout America and eventually the rest of the world.

Toaster

Toast comes from the Latin word *tostare* meaning 'to brown by heat'. The Romans impaled bread on an implement and browned it in a flame. Toast wasn't so much an alternative to bread but rather it was made to make the consumption of stale bread less unpleasant an experience. Early electric toasters appeared in the late 1800s, made by the Crompton company in the North East of England after the discovery that electric elements could be fashioned out of nichrome (a nickel–chromium alloy) due to its high heat resistance over extended periods of time. This was all very well but, with a good working element, there was nothing to stop the toaster from just keeping on toasting – if one walked away and forgot about it, the toaster was prone to starting fires. In 1919 Charles Strite decided to address the burning issue of toasters that didn't stop toasting (which was a particular problem for those in the busy catering industry). He designed the first toaster complete with a timer and a mechanism to eject the toast after a set toasting period. His pop-up toaster was patented and made available to the general public in 1926.

Torch

The first electric torch was made in America by the Bristol Electric Lamp Company in 1891. Weighing in at 2lb (battery included), the first significant order for the square 'Bull's Eye Lantern' came from the Bristol General Omnibus Company. They bought 60 and gave them to their ticket inspectors, thus making it harder for people to ride without a ticket! Ever Ready made the first British electric torch. Their Electric Torch No. 1 used batteries that were said to give

'5,000 flashes'. Moreover, advertisers were keen to stress that the torch could be turned on inside a barrel of gunpowder without causing a problem. Not, mind you, that they were recommending it as a practice.

Tupperware

Just think: once upon a time there were no plastic containers around in which to stow one's sandwiches or pasta salad. Well, Earl Silas Tupper realised the awfulness of this situation before anyone else and resolved to sort it out. An American chemist, he did much to further the development of new plastics with his Tupper Plastics Company, formed in 1938. Using hard polyethylene slag, a by-product of the oil-refining process, he created excellent mouldable plastic and began making his famous airtight food-containers in 1945. Bizarrely, one might think, Tupperware didn't immediately rock the average American housewife's world. Door-to-door sales met with some success, but it was sales rep Brownie Wise who sexed everything up with the concept of the Tupperware party. The advent of these riveting occasions sealed the company's good fortune – Tupperware is now sold in 100 countries around the world, and it is claimed that every two minutes another Tupperware party gets going somewhere on the planet!

Vacuum Cleaner

The vacuum cleaner is perhaps the only invention whose history begins with a handkerchief stuffed in a man's mouth. The mouth belonged to British engineer Hubert Booth who, in 1902, sucked at a chair's upholstery through his hanky in order to confirm to himself that, when it came to collecting dust, sucking was better than blowing (he was considering the merits of an early device that only displaced dust temporarily by blowing it from the surface it had settled on). Sure enough, his handkerchief was filthy once he'd finished. He then set about putting together the world's first vacuum cleaner, and the result was a device so large it could not be taken into buildings! The huge motor and pump had to remain on a horse-drawn carriage in the street while a hose nearly 250 metres long was taken into houses and offices. Though admired by King Edward VII, Booth's vacuum business eventually faded, which 'sucked' a little.

The man responsible for the development of the domestic vacuum cleaner was James Spangler. A janitor who suffered from nasty asthma, he pined for a cleaning method that got rid of the dust that brought on his attacks, and invented a machine to do just that. The idea was simple: he took an electric fan and connected it to a pillowcase, which was connected to a broom handle and a rotating fan. The vacuum created caused dust-filled air to be pushed into the bag. Spangler had created the first upright vacuum cleaner. Soon enough his cousin's husband, William Henry Hoover, realised the huge commercial potential of such an appliance and promptly purchased the rights in 1908. That same year, the first Hoover went on the market, and the company went from strength to strength. A household name in Britain, such is the ubiquity of the brand that people use the word 'hoover' to refer to any brand of vacuum cleaner, and 'to hoover' is used as a verb.

The basic principles of the vacuum cleaner remained fundamentally the same for many decades until James Dyson invented the cyclonic vacuum cleaner. Aware of the problem of clogged-up bags causing reduced suction in vacuum cleaners, and inspired by an industrial cyclone – a device used to remove dust from air leaving a factory – he felt that a bagless, and more efficient, vacuum cleaner must be possible. He was right, and from the moment of his initial idea it took 15 years and 5,000 prototypes before a final design was arrived at. The first Dyson was produced in 1993 and they are now the most popular vacuum cleaners in Europe.

CHAPTER THREE:
FOOD

'No can do !'

Can Opener

FOOD

Baguette

A diverting myth about the development of the baguette states that it was invented by Napoleon when fighting the Russians. Due to cold Russian temperatures, his soldiers were ordered to pack extra clothes, which meant there was less space for food. The baguette was conceived to get around this problem – its shape meant it could be stowed down the trouser leg of the chilly soldier. A fine tale: if only it were true. Unfortunately, the fact that the Napoleonic army travelled with mobile bakery units, and that the soldiers' uniforms were prohibitively tight, puts paid to such a theory. However, the true origin is itself interesting.

Baguettes descended from Viennese bread in the 1800s. Steam ovens were a new arrival, and this method of cooking allowed for a perfectly crisp crust and soft centre that typifies the baguette. The long and slim style of loaf became more and more prevalent because it required less cooking time than traditional round loaves. The need for quick-bake loaves arose from a 1920s law prohibiting bakers working before 4am.

Baked Beans

The recipe for baked beans is quite probably based on a Native American preparation where beans were cooked with bear fat and maple syrup. Settlers in America adapted this recipe and used pork fat and molasses instead. In the mid-19th century, over half a century before the baked beans we know came about, this way of preparing beans was very popular and, in New England, a 'baked bean festival' used to take place on Saturday nights. After the beans were soaked overnight, mustard, pork and molasses were added to them before they were carried through the streets at dawn by local children. Pots of beans were baked in ovens all day and feasted on by the locals that night. A canned variety of the same recipe turned up in 1875, when the Burnham and Morrill

Company of America began producing them for the fishermen who were sore that they often missed out on the Saturday-night bean-fest.

What we recognise as baked beans – those covered in tomato sauce that is – were first produced in America in 1895, and in Great Britain it was the American company HJ Heinz that tested them out on the Brits. Marketed to the working classes as a cheap and nourishing feed, they were an instant hit. By 1928, manufacture had moved from the US to Great Britain, and these days over a million cans of beans are consumed every day in the UK.

Biscuit

Where would we be without tea and biscuits, or 'cookies', as they're known in America? A little slimmer perhaps? Who knows. What we do know is that the word biscuit comes from the Latin *bis cotum*, which means 'twice cooked', and that 'biscuits of muslin' were taken along on crusades by Richard the Lionheart, although it's uncertain whether or not he had a nice cup of tea into which to dip them. Another historically significant biscuit was the British naval hard tack, made from flour, water and salt and used on long voyages in the 17th century because they were so much longer-lasting than bread. Spreading to

America via British and Dutch immigrants, hard tacks proved very popular and were produced in infinite varieties in the following centuries.

In 1937 the biscuit scene was transformed by the arrival of the chocolate chip cookie. Feeling a little creative, American housewife Ruth Wakefield added some chopped-up chocolate to her cookie mixture, naturally expecting it to melt when cooked. How wrong she was! To her utter amazement, the chocolate remained pretty much intact. Unwittingly, Ruth had created a biscuit that would have made Richard the Lionheart salivate in his grave. Mrs Wakefield's contribution to the great American milk and cookies ritual should never be underestimated!

In England, it may come as a surprise that biscuits were quite recently at the centre of a stormy court battle between the Inland Revenue and McVities, the company that produces Jaffa Cakes. The tax men claimed that the little biscuit-sized cakes were in fact chocolate-covered 'luxury' biscuits and that VAT was therefore payable on them (while biscuits and cakes are 'zero rated' for tax, chocolate-covered biscuits are eligible for VAT). However, McVities pointed out that 'biscuits go soft when stale, whereas cakes go hard when stale'. They won their case, and probably relaxed afterwards with tea and biscuits, or perhaps cake …

Canned Food

The way to an army's heart is surely through its stomach. In 1795, when Napoleon became concerned that large amounts of food were rotting before his troops got to eat it, he promised an award of 12,000 francs to anyone who could invent a way to preserve the meat and vegetables for his boys. Nicolas Appert, a sweetshop owner in Paris, came closest to answering Napoleon's call by making a sterile, vacuum-sealed container within which the food could be contained, though he didn't know exactly why it kept it fresh.

Though Appert's container kept the food fresh, it was made out of glass, which meant it was too fragile to carry on long journeys. In 1810 Peter Durand, a British inventor, made cans from steel and lead, but the result was said to be responsible for a number of deaths. In 1845 Sir John Franklin's ship set off with 130 crewmen from England in search of the Northwest Passage. With a piano and hundreds of books for entertainment, in addition Franklin stowed supplies of canned food that could potentially feed his crew for three years. They never returned, and one theory states that they were poisoned to death by the lead from the seals. Tin-coated steel cans, still used today, solved this problem, but canning processes remained laborious and the times required to thoroughly cook food to be contained in the can made the whole enterprise very lengthy and costly. With improved mechanisation, these problems were eased and canned food became available to more of the population. Demand for food that was high in calorific value as well as being long-lasting increased in the Western world, especially during the wars of the late 19th and early 20th centuries. Canned food fitted the bill perfectly, and after the war new varieties were added by companies such as Heinz. Additionally, on the marketing side, can labels were made increasingly more attractive to give the impression of something wonderful within!

Can-opener

It is perhaps rather hard to believe that the can-opener was invented nearly 30 years after food canning was perfected! Before this, cans had to be cut with bayonets, pierced or smashed against something hard – they were even shot at! Once thin metal was available for cans, the first can-opener, which drove a curved blade into the metal, was invented by Robert Yeates in 1855, but it wasn't patented until the late fifties by American Ezra Warner.

Cheese

You don't need ingredients to make cheese – you need *one* ingredient: milk. Milk curdles to make cheese due to a natural enzyme, rennin. It is this simple process that has enabled production of cheese for so many years. About 5,000 years, in fact. Sumerian cave paintings from that era show cheese being prepared in containers. It is supposed that a Sumerian herdsman must have first discovered it by chancing on a bit of cheese in one of the calf stomachs that were used at the time for storing milk. We can only pity the poor fellow that he didn't have any biscuits to go with it …

Cheese is mentioned several times in the Bible, and gets a mention in Homer's *Odyssey*. In Roman times cheese was a common, everyday food, and the cheese-making process was considered an art. The method used was not very different from that of today. It involved rennet coagulation, pressing of the curd, salting and aging. Pliny's *Historia Naturalis* devotes a chapter to the plethora of cheeses enjoyed by Romans of the early Empire.

Cornish Pasty

The precise origins of the pasty are obscure, but the earliest known written recipe is dated 1746 and can be seen at the Cornwall Record Office. Pasties developed from the needs of Cornish miners who had to spend long periods underground without returning to ground level for lunch or dinner. The thick crust was not designed to be eaten, but rather it was there so the miners could hold it with their dirty hands and eat the nutritious filling, before discarding the dirty pastry afterwards – the waste pastry was thought to please the 'knockers', members of the spirit world who had the ability to bring disaster to a mine if not kept happy. The thick pastry's other functions were to insulate the food within it and keep the miners warm when it was stored close to the body. There were compartments within each pasty for meat, vegetable and even a bit of pudding (apple pie was popular), and mine-town bakers would personalise each one, making bespoke pasties to order and identifying which miner they were meant for with his initials carved into it, or raised up in the pastry. Pasty tradition states that one measure of a good pasty is whether or not it can survive falling down a mine shaft, although nowadays pasties are a popular fast food, more likely to be seen falling from an office worker's desk, or a commuter's hand as they fumble for their mobile phone at the train station.

Crisps

In 1853 George Crum was knocking out fried potato chips with every dish in his restaurant in Saratoga Springs, New York. One day he ended up serving something quite different, a variation on the potato chip, because a grumpy regular, Cornelius Vanderbilt, complained he was fed up that the chips were too thickly cut, soggy and 'not a nice mouthful'. Irritated, Crum set about making a potato chip that was so thin it couldn't easily be fried in a pan, nor picked up with a fork. Gleefully he served a plate full of his absurdly thin potato chips to Vanderbilt. But Vanderbilt loved them and, accordingly, Crum added them straight to the menu To Crum's surprise they were a huge success. The new crisp not only became a popular choice in the restaurant; word spread so quickly that the starchy little snack was named the Saratoga Chip. Chip production began in domestic kitchens with family-run businesses distributing them all over the USA.

Initially, barrels were used for long-term storage, but chips quickly became bland and stale, particularly at the bottom of the pile. American entrepreneur Laura Scudder persisted in finding a cheap, light container that would seal the snack for ultimate crispness. Scudder made bags out of wax paper and heated them to form airtight seals. Nitrogen was injected to provide a barrier to prevent chips from being crushed and also to extend the shelf life, and this was the first chip packet.

A man known only as Carter is reported to have taken the idea of the chip from France to Britain and made them commercially available. But he was usurped by Smiths, who began mass production in 1920. Smiths are still going strong today. And, to be clear, the British call them crisps!

French Fries

Depending on who you speak to, the French fry originated in either France or Belgium. On the one hand, the French are adamant that it is their invention, and back their claim up with the fact that, writing from the early 1800s, US President Thomas Jefferson mentions 'potatoes deep-fried while raw', and that he got the a recipe from his French chef, Honoré Julien.

On the other hand, Belgian historian Jo Gerard recounts that potatoes were already fried in 1680, in the area of 'the Meuse valley between Dinant and Liège, Belgium. The poor inhabitants of this region allegedly had the custom of accompanying their meals with small fried fish, but when the river was frozen and they were unable to fish, they cut potatoes lengthwise and fried them in oil to accompany their meals.' In addition to this, the Belgians sometimes claim that the term 'French' was applied to the fries because British or American soldiers in Belgium during the First World War ate Belgian fries and called them 'French' because it was the official language of the Belgian army at that time. But the term 'French fried potatoes' had been in use long before the war. At the end of the day, it's a little bit of a mystery which nation came up with the dish, and slightly amusing that both are so precious about a foodstuff that most of us now associate with cheap fast-food outlets all over the world.

Frozen Food

Some say we have Sir Francis Bacon to thank for making the first tentative steps towards freezing food. His early experiments involved stuffing chickens with snow on Highgate Hill in the 1500s. Unfortunately his method wasn't successful.

Clarence Birdseye is more widely accepted as the

man who perfected the preservation of food by freezing and making the method commercially available. It was while studying in the Arctic that old Clarence observed Eskimos preserving their fish for months on end by dunking it in freezing holes in the ice straight after it had been caught. Attempting to emulate these freezing conditions back at home, Birdseye tried various methods including the use of crushed ice, with added salt to further lower the temperature. But his success was limited, because when the food defrosted it fell to bits. Thinking back to the Eskimos' fish-freezing methods, Birdseye worked out that an important factor is the speed at which food is frozen. Today we know that freezing food slowly causes it to lyse (the cell wall ruptures and ice crystals can get into the food, burning and destroying its structure and flavour). Birdseye realised that simply 'snap freezing' at very low temperatures (around -40°C) is the key.

In 1924 he developed a quick-freeze system where food would be wrapped in waxed boxes, the way it would be presented to the customer, and snap frozen under high pressure. At first he used this for the sale of fish, but then widened his captain's net to include vegetables and fruit, finding that berries froze particularly well, retaining a high proportion of their nutritional value. The name Birdseye was modified to Birds-eye, and the frozen-food world really got exciting when Birds-eye ready meals first hit the shelves, hailing the arrival of the frozen TV dinner!

Hamburger

The 'ham' in 'hamburger' has nothing to do with pork meat. Rather, the name comes from Hamburg in Germany, in much the same way that frankfurters come from Frankfurt. But hamburgers were not invented in Hamburg – the name was merely given to the dish once it arrived in Germany from other shores and established itself. Ground-beef cakes existed long before the term hamburger was coined.

Thirteenth-century Mongol horsemen would store raw minced meat under their saddles while running riot in central Asia, thus tenderising the meat while at the same time giving them an easily accessible source of food – no need to stop for lunch when doing a spot of invading if you're sitting on it! On arrival in Moscow, the Mongols showed off their special food and the Russians called it 'steak tartare', the Mongols being known as 'Tartars'. Later on, German ships from the Port of Hamburg trading with Russian ports brought the concept back home with them, and the hamburger was born. Variations of it spread throughout Germany and were eventually taken to America by German immigrants and traders. Once there, the modern hamburger in a bun evolved throughout the early 20th century, though who produced the first of them is hotly disputed to this day.

Hot Dog

According to the National Hot Dog and Sausage Council, the exact origin of the hotdog is not known. Various theories exist, though. One is that a 17th-century German butcher named Johann Georghehner was the creator of the snack. As for the term 'hot dog', it is often attributed to a New York cartoonist who annotated his sketch with the term. The sketch depicted people munching away on sausages in long bread buns from a snack wagon at a local baseball stadium in 1902. The wagon vendor was selling German sausages called Wieners. The word 'Wiener' was something Americans associated with a dog due to the Wiener dog being the American name for the Dachshund. The phrase coined by the cartoonist is said to have stuck for the popular snack, still served at sporting events today. A 'hot-dog historian' quoted on the National Hot Dog and Sausage Council website claims the term was first circulated on campus at Yale University – at first the students started referring to the vehicle where the hot snacks were served from as 'dog wagons', and the term 'hot dog' is said to have evolved soon after.

Margarine

In 1869 Napoleon III publicised that he would handsomely reward the first person who could invent a butter substitute. He wanted something cheaper than butter that could be made available to the working classes. The French chemist Hippolyte Mège-Mouriés came up with the mouth-watering combination of beef fat mixed with pig stomach. He named it oleomargarine, presumably a hybridisation of margaric acid and oleic acid, which are two fatty acids contained in animal fats. Despite the obvious appeal of this new delicacy, the 'spread' of margarine use was hampered hugely due to protestations from the dairy industry in the USA, who felt threatened by a cheap imitation. Particularly contentious was the fact that margarine producers were dying their naturally white product yellow to further align it with butter. The butter war meant that by the early 1900s American laws prevented eight out of ten people being able to buy margarine and the few sales that did go ahead were heavily taxed. In Australia, would-be marge-spreaders had to wait until the 1960s before they could legally purchase coloured margarine! Further improvements and modifications continued to be made to the basic margarine recipe, including the addition of hydrogen to make a more hard-setting spread. Nowadays margarine is a generic term for many oily spreads.

Mayonnaise

Various explanations exist for the origins of mayonnaise. One story is that it was invented for the Duke of Mayenne in 1859. Another popularly quoted story is that it was created by a chef working for the Duc de Richelieu in 1756 when the French were victorious over the British in seizing Mahon, the capital city of Minorca. The chef took the dressing to a celebratory feast as a special treat because it was especially rich and luxurious in texture. Mayon is said to have come from the name Mahon and 'aise' is the French suffix for something coming from a place.

Mustard

Mustard is a member of the Cruciferae family of plants and grows all year round in Europe, North America and England. Though it is sometimes grown solely for the sake of ploughing it back into the earth to enrich surrounding soil, its most popular use is as a tasty condiment and a most welcome guest at mealtimes. Mustard wasn't always as pure and full flavoured as it is today though. Aficionados have refined the recipe and fought to strictly limit and regulate the number of mustard manufacturers in order to control which ingredients are added. The general recipe involves crushing the naturally aromatic seed and sieving it to remove the husks, leaving the fine mustard flour. The flour is mixed with water, vinegar or alcohol to form a paste and further flavoured with herbs and spices. Although earlier mustard-making methods were documented in AD 42, it wasn't until the 10th century that French monks made the recipe official and began venturing into commercial enterprise. This caught on in Paris and became something of a favourite item for sale by street vendors. But laws became increasingly strict, with 'clean' laws imposed making it an offence for anyone without a licence to produce mustard. Mustard laws slackened, and by the mid-1850s a good bit of healthy competition forced recipes to be optimised and improved. Dijon became the capital's favourite, and the recipe involved removing vinegar for a smoother flavour.

However, it was Jeremiah Colman from Stoke in the UK who in 1814 came up with a unique product that remains a household icon today. Colman's bright-yellow tin contained a special formula of crushed brown seeds (*Brassica juncea*) combined with crushed white seeds (*Sinapis alba*) to produce a bright-yellow powder. It ended up being a money-spinner for him. The saying goes that Colman 'made his money by what people left on their plate', because people always take more out of the jar than they need and, once out, it doesn't go back in. The only option, of course, is to buy more.

NutraSweet/Artificial Sweetener
Despite endless reports that excess sugar is harmful to our health, some of us just can't shake the habit of two sugars in our tea. So it was fortunate when this special molecule was accidentally discovered to be sweet by chemist James Schlatter in 1965. The search had been on at Searle & Co for a drug to combat stomach ulcers, with Schlatter working on the synthesis of aspartame. Mid-experiment, he made the mistake that all lab scientists know to avoid – he licked his fingers! To his great surprise he tasted something pleasantly sweet, even sweeter than sugar itself. Searle organised for Aspartame to

go though the necessary stages of testing, but it would be a number of years before the FDA approved it, for fears had surfaced that it could cause brain cancer. Eventually it was passed for use in dried foods and in the early 1980s the FDA released approval for Aspartame to be added to fizzy drinks, cakes and other foods. In 1995 Aspartame began being sold under the brand name NutraSweet, which remains a household name in a weight-conscious world.

Pizza

Written records show that vegetables flavoured with herbs and spices on unleavened bread were eaten by the Ancient Greeks, but the pizza that we know today came out of a visit by royals to Naples, a visit that ended up making it the true home of pizza. Raffaele Esposito, an Italian pizzaioli (pizza chef), wanted to mark Queen Margarita's visit to Naples in 1889. He picked the pizza as it was a popular local dish. Over the base he spread a tomato base followed by a covering of mozzarella and basil to symbolise the three colours of the Italian flag. Esposito named it 'Margarita' in honour of the queen, who had thanked him for the dish. Up until this time the tomato base had not been a feature on the pizza – it was a great success and the idea was soon exported to America. After the Second World War, it spread around the world and is now the basis of a large majority of high-street restaurants. The word 'pizza' comes from the Latin word *pinsa*, meaning flatbread. A legend suggests that Roman soldiers developed a liking for Jewish matzoth while stationed in Roman-occupied Palestine and developed a similar food after returning home.

Quorn

Quorn may never have come about were it not for a panic that turned out to be a bit of a fuss over nothing. In the 1960s, scientists predicted that the world would soon be short of enough protein to feed our animals, and therefore also short of enough animals to feed us. Consequently, ICI employed Englishman Dr Peter King to come up with a scientifically produced animal-product replacement. This he did, and called it Pruteen, but alas the shortage never came about so it seemed there would be no need for it. But thanks to vegetarianism, it turned out their work would be put to use after all. ICI got together with another company, Rank Hovis McDougall, who had been developing mycoprotein (a naturally occurring fungus that had been discovered in the late 1960s and found to be perfectly edible and very nutritious) and helped them successfully mass-produce it. After many years of testing, permission was given to sell mycoprotein in 1980, and it was launched in 1986 as Quorn. It was very popular, but the supermarkets did not want to know until Lord Sainsbury agreed to stock it in his British supermarket chain, Sainsbury's, in the early 1990s. Other chains followed suit. In 2004, McDonald's introduced a Quorn burger, replete with a seal of approval from the Vegetarian Society, and today around 60 per cent of meat-replacement foods sold in the UK are Quorn products, with annual sales totting up to roughly £95 million!

Sandwich
John Montagu, the fourth Earl of Sandwich, is often credited with having invented the sandwich in 1762 when he asked one of his servants to put some cold beef in between two slices of bread as he did not want to leave the gambling table he was playing at to attend dinner. In fact, this way of eating food was already common – what came out of the event was merely that food between bread was henceforth referred to as a sandwich. Edward Gibbon, the 18th-century English historian and

member of parliament, was the first to refer to sandwiches in his journal.

There are many ways of referring to a sandwich – among them 'butty', 'piece', 'sarnie', 'sanger' and 'sub'. But, whatever we like to call them, we certainly love to eat them: in Britain around 1.8 billion sandwiches are purchased outside the home every year!

Sausage

Early sausages were made of goat meat in China around about 3000 BC. In Ancient Greece, a type of roasted blood sausage was eaten, and the Romans enjoyed a sausage or two for many years before they transported the idea to Britain. The name sausage comes from the Latin *salcisius*, which means salted meat. Salting was a way of preserving meat before refrigeration was possible. What was special about the invention of the sausage was that it avoided wastefulness. It made use of meat offcuts and bits of offal once the prime cuts had been removed from the carcass. Meat scraps were never going to be a delicacy, but making sausages cleverly disguised the appearance and overall taste of the nasty bits by homogenising the flesh and adding herbs and spices. Unfortunately, one couldn't always be certain of what 'secret' ingredients were contained in the sausage – various unsavoury things were used to bulk out the mixture. In Britain, sawdust was one of the more unappetising additives used in the early 1900s. Sausage skins used to be made from animal intestine (another efficient use of animal waste), but nowadays they are commonly made from non-animal-based materials such as cellulose. Today's sausages don't contain sawdust either – breadcrumbs are usually used as fillers and the meat and fat content is regulated. In the USA fat content must be no more than 30–50 per cent, depending on the particular type of sausage. The slang name for sausage – the 'banger' – came into common use because at one time they were so cheaply manufactured that the high water content caused them to pop and bang when they were being cooked.

Sliced Bread
The invention of sliced bread is a story of determination against the odds. Having filed his first patent for a bread-slicing machine in 1912, it took American Otto Frederick Rohwedder no less than 17 years to find commercial success with his invention. To cut, or perhaps slice, a long story short, poor Otto was plagued by problems from the off. His first hurdle was the resistance he encountered from the

bakers of the time, who firmly believed that sliced bread was a terrible idea because it would go stale quickly. No problem, reasoned Otto, and invented a machine that not only sliced the bread, but also wrapped it. Problem solved, you may think, but in 1915 he went down with a kidney disease that would keep him out of action for seven years. To add insult to injury, in the midst of his illness, Otto's factory and all of his prototypes burned down. But, like good bread, Otto rose to the occasion, recovered from his illness and set about re-establishing himself. It was not until 1928 that the Chillicothe Baking Company of Missouri took a chance on his idea and began selling bread that had been sliced the Rohwedder way. Five years later, only 20 per cent of the bread sold in America was unsliced.

Supermarket

The supermarket evolved from the traditional grocery store – an outlet where the customer would have very little choice between brands, and items had to be measured and packaged separately. The grocer was required to retrieve everything a customer needed, and there would only be one till. This made the old-fashioned weekly shop a time-consuming affair with a lot of waiting around involved. The modern-day supermarket not only offers a staggering array of food choice, but also increasingly a wide range of other products and services such as insurance services, clothes and furniture, and even banks.

The grocery store where customers first enjoyed the novelty of serving themselves was set up in California in 1912. After the First World War, more of the characteristics of today's supermarkets developed in a chain of stores in Tennessee. At Piggly Wiggly chain stores, turnstiles were in place, plenty of room was made to ensure customers had easy access to the shelves so that they would serve themselves, and several tills were in operation to keep things ticking over quickly.

Michael Cullen is said to be the first person to take the Piggly Wiggly concept to the next level. In 1930 on Long Island, Cullen opened up a store that was 12,000 square feet. Revenue rose so quickly that, thanks to economies of scale, he was able to offer the lowest prices by far, which made his shops even more popular with customers spending what little money they had during the Depression.

In Britain, Jack Cohen had a grocery store in the East End of London. On a trip to America in the 1920s, he was impressed by the concept of the supermarket over there. He brought the idea back to England and opened up his grocery store to the self-service layout. He named it Tesco, a hybrid of his wife's name, Tessa, and their surname Cohen. By 1939, 100 Tesco stores had their tills ringing. Today, Tesco is the most successful supermarket in Britain.

CHAPTER FOUR:
DRINK

'We'd better drop the Dom Perignon or
we'll never launch the thing !'

Champagne

DRINK

Alcohol

It seems that binge drinking isn't quite the modern phenomenon we're led to believe it is, as some of the first laws in the history of mankind relate to Babylonian drinking houses. The laws were created in 1770 BC, which suggests that even then people loved to drink, and that they didn't like to stop at one! As far as the discovery of alcohol and its intoxicating power goes, it is believed that man would have first got tipsy from alcohol about 7,000 years ago by unwittingly consuming it in the form of fermented fruit or honey.

Beer

Beer hasn't always been served in pints. In fact, it has been around for thousands of years. It was first brewed in Mesopotamia, and then Egypt. It was produced from barley, or other grains, steeped in water, and the sugars derived from it were fermented with yeast. This are the essentials of beer production, full stop!

Brandy

Brandy's arrival on to the booze scene was the result of a happy accident. It became a real hit around the 15th century BC, though the beginnings of distilling alcohol were charted much earlier. Wine was concentrated to reduce its volume for transportation with the intention to dilute it back to the concentration of a standard wine at the end of the journey – some say the volume was lowered in order to reduce taxes paid en route on a tax-by-volume system, others say it was to lessen the load carried on ships by Dutch merchants. Either way, when the barrels were opened and the 'wine' was sampled, a rather different beverage was discovered. What came out was different in flavour and chemical structure to what had been put in. Rather than weakening it with water, it was swiftly enjoyed.

Brandy takes its name from the Dutch word *brandewijn*, meaning distilled or burned.

Carbonated Drinks

Given the bad press that fizzy drinks get these days, it may come as a surprise that 'soda pop' was originally conceived as a beverage to be consumed for the benefit of one's health. Artificially carbonated water was first created in 1741, but it was Swiss-German Jacob Schweppe who first mass-produced what were called 'aerated waters' in the 1780s. From this emerged the concept of flavoured fizzy water, pioneered in America around the turn of the 19th century. During this time, pharmacists were keen to capitalise on the restorative qualities of mineral water by adding a range of health-giving ingredients to them. They experimented with dandelions, ginger, sasparilla and even birch bark, but the most famous ingredients to be used were coca and kola, which of course lead to the most famous soft drink of them all, Coca-Cola (see page 87), first sold in 1886. Along with other concoctions, initially it was sold at pharmacies as a health tonic. Eventually the soda fountains used to dispense it were replaced with bottles able to hold the fizz in until opened, which happened towards the end of the 1900s.

Champagne

It was said that Marylyn Monroe and Jayne Mansfield loved to bathe in champagne, which must have been lovely for them, but most of us have to settle for a glass or two now and then, unless of course we're sending a boat on its maiden voyage in which case we might have the good sense to chuck a bottle of the stuff at its hull. Whatever we do with champagne, we're probably too sozzled to bother questioning who invented it or how the fizz that pops the cork out of the bottle gets in there in the first place. Well, it is the carbon monoxide in the wine that makes it so bubbly, and it is created thanks to a unique fermentation method called *méthode champenoise*. Its inventors were monks from the Champagne region of north-eastern France, and brothers Jean Oudart and Dom Pierre Pérignon are usually given credit for first producing it around 1670. However, the chances are that many winemakers in the region were at the same game, and through modern dating methods available it has been established that sparkling wine was in production as early as 1535. So who to raise a glass to for the creation of champers is a tough call, but we do know that Dom Pérignon came up with the wire collar to hold the cork in place against the building pressure during fermentation, so three cheers to him!

Coca-Cola

Coca-Cola may be frowned upon by some as an unhealthy drink these days, but at one time it was seen as a healthy alternative to alcohol. A pharmacist named John Pemberton invented something called cocawine, a cola and wine mix adapted from a European drink. In a clever response to the Prohibition in 1895, Pemberton adapted cocawine and made a carbonated non-alcoholic drink. However, the ingredients contained didn't make it so innocuous. In addition to a mass of sugar, it contained cocaine from the leaves of the coca plant. He came up with a name, a hybridisation of the two and called it coca-cola. By 1892 it sold so well that he was able to form the Coca-Cola Company. In 1903, the active form of cocaine was removed due to its highly addictive nature. At first Coca-Cola was only available from soda fountains at local pharmacies, and it was marketed as a medicinal drink to cure numerous ailments, from headaches to morphine dependency.

Cocktail

An article in the American magazine *The Balance* describes cocktails for the first time in 1806 as being 'a kind of stimulating liquor composed of spirits of any kind, sugar, water and bitters'. The origin of cocktails is fairly well contested, but the most oft-repeated story is that US bartender Betsy Flanagan had been decorating sophisticated drinks she was serving with the feathers from her chickens. A drunk local requested another drink by shouting a request for 'one of those cocktails'.

Coffee

Around AD 1000, a goatsman named Kaldi observed his herd becoming much more energetic after munching on the berries of the coffee plant – *Coffea arabica*. Intrigued, Kaldi tried a few himself and experienced the energy buzz we're all familiar with. Amazed by the invigorating power of the beans, he spread the word and before long everyone was at it.

But it was the Ottoman Turks who truly popularised coffee as a drink. They imported it to Constantinople in the 1400s and opened Kiva Han, the first recorded coffee shop.

In 1930 Max Morgnethaler, a Nestlé employee, searched for ways to make a coffee that would be quick and easy yet still capture the flavour of coffee poured from a percolator. Prior to this, many rather unsuccessful attempts to develop instant coffee had been made. First came distilled coffee essence in liquid form, which failed because, quite simply, it couldn't retain the flavour of coffee. Later attempts included grinding the beans to an extra fine powder in the hope that it would dissolve, but this was a failure too. Rather than dissolving, the ground coffee only formed a thick film on the surface of the hot water. Something cleverer had to be done. The coffee needed to be fully dissolvable. In 1938, Nestlé embraced the idea of spray-drying. This involved spraying fresh coffee into the top of a tower where a large supply of hot air would instantly dry it into droplets. The coffee droplets were then rendered into granules. Spray-drying was an improvement on previous methods, but in the early 1960s Nestlé began selling a further-improved instant coffee that we still enjoy today – freeze-dried instant coffee. Freeze-drying is a sophisticated, multistage technique used in many fields such as pharmacy. It was found that coffee could be frozen into thin sheets, which could then be ground into particles. Once subjected to a vacuum, these particles left granules of pure, dissolvable coffee, flavour intact.

Nescafé produced Blend 37 – the first instant freeze-dried coffee – in 1923. But why '37'? According to Nescafé, it's so named because racing driver Didier Cambresson drank nothing but Nescafé coffee to keep him going during the gruelling 24-hour Le Mans motor race. Driving solo because his race partner didn't turn up, he came 37th in car 37.

Drinking Straw

The story goes that straws were fashioned in ancient times to facilitate drinking crudely made beer containing unpleasant floating solids resulting from the fermentation process – the straw stopped the solids getting into one's mouth. However, it is Marvin Stone to whom we should be thankful for patenting the hugely successful spiral-effect design that is still used today. Employees at Stone's factory were producing cigarette holders out of paper at the time. It was when Stone thought about the phenomenon of people sucking their beverages up through hollow strands of grass that he decided to emulate the 'devices' with paper and make them more widely available. He adapted the dimensions of the cigarette holder but, knowing that paper would fail because it wasn't waterproof, he tried using paper with a paraffin finish. Then he worked on the difficult part – fashioning the paper into a long thin cylindrical shape and being able to mass-produce it efficiently. Marvin was delighted when he realised that simply winding the paper carefully and tightly round a cylindrical rod of about ½cm in diameter would do the trick. He glued it in place and there he had it, a secure and waterproof straw. Stone eventually designed machines to carry out the process, and in 1888 he patented the drinking straw, protecting the 'winding' design because he thought the principle may be useful for things other than the consumption of fizzy pop. He was right. In the early part of the 20th century, the 'winding' design began to be used industrially and the tubes were required in many fields ranging from engine construction to surgical equipment.

Espresso

Quintessentially Italian, the single black espresso shot forms the basis of any grande-decaf-skinny-soya-mocha-type fancy business one may find in today's chain coffee houses. As suggested by the name, espresso came about from one Italian engineer Achille Gaggia's need for a speedy coffee hit. His 1938 spring piston lever machine pumped not quite boiling water through ground coffee very quickly, eliminating the opportunity for the bitter taste to contaminate it. One pull of the lever ensured a perfect, quick coffee and, once patented, the espresso machine became beloved of café owners all over Italy, and then the world.

Gin

Gin was invented in the Netherlands by professor of medicine Franciscus Sylvius at the University of Leiden in the 17th century, and the word derives from the Dutch *jeneverbes* for 'juniper', which in turn comes from the Latin *juniperis*. Attempting to fabricate a cheap alternative to juniper berry oil, Franciscus used a distillation of the berries with spirits. Originally intended for use in medicine as a diuretic, it quickly became used for leisure!

Gin came to England with William of Orange, the Dutchman who sat on the British Throne in the 17th century. It was hugely popular in England, particularly once unlicensed gin production was legalised and heavy taxes were imposed on imported spirits. England's notorious 'gin shops' sprung up everywhere, and the nation's huge consumption of the stuff was blamed for many social ills.

Gin and tonic originated in tropical English colonies where gin came in handy for hiding the unpleasant taste of quinine, a substance used to protect against malaria. Quinine was dissolved in fizzy water to produce tonic, and long after quinine ceased to be used for malaria prevention, the G and T survived as a protection against sobriety!

Milkshake

One invention can often lead straight to another. This was the case with the modern milkshake, which came about as a result of the blender introduced by Steven Poplawski in 1922. Earlier milkshakes were made as nutritional supplements for those with a variety of ailments, and William Horlick contributed his own delightful twist by adding malt to create the malted milkshake.

Pasteurised Milk

Up until the time of Louis Pasteur, a 19th-century French microbiologist, it was difficult to keep a readily available supply of fresh milk. In response to a request from a distiller in Lille, Pasteur was experimenting with ways in which to prevent alcohol from souring during fermentation. The fermentation process was little understood at the time, and was popularly believed to be the result of a chemical reaction, but Pasteur's microbiological expertise led him to reveal this wasn't the case. He knew that, if fermentation came about from a chemical reaction, the product would be made up of a 50:50 ratio of two structural forms known as enantiomers, one a mirror image of the other. A number of tests, and the study of the soured distillate under the microscope, led him to realise that only one enantiomer was present. This was an effect Pasteur knew could only have come about from the presence of micro-organisms. Pasteur worked out that yeast was a living organism and that it was this that was causing the souring. He set about trying to find conditions that would kill the yeast without destroying anything else in the liquid. He discovered that heating it to 65°C for 30 minutes, followed by rapid cooling, was the answer to reducing the number of disease-causing micro-organisms contained in the liquid, and that this process would keep it fresh. This was applied in industry, one of the first uses being the prevention of souring in alcohol. It was later applied to milk and the process became known as pasteurisation in recognition of the scientist's work.

Ring Pull

Remember the time when a ring pull used to separate entirely from your can of Coke. It seems like an age away. Further away still are the 1930s, when cans of drink needed to be pierced twice with a separate 'church key', which was fine as long as you didn't leave your church key at home and find yourself out and about with nothing but thirst and a sealed can to keep you company. In 1959 American Ermal Fraze found himself in precisely this situation at a picnic – reduced to using the bumper of his car to pry open his can of beer, he resolved to find a more convenient way of getting to the good stuff. Thinking cap on, he came up with an idea for a built-in lever that could lift up and push open a section of scored metal on top of a can. He patented his idea in 1963, but his solid lever tab was a bit tricky to get one's fingers under. American inventors Omar Brown and Don Peters improved the design in 1965 by adding the vital ring-pull design, and forever more America, and the world, found it infinitely easier to get at their sugary beverages. The one remaining problem was the sharp ring pulls that ended up as litter and cutting people's feet on beaches – Omar Brown came to the rescue again in 1973 with his design for the 'push-in fold-back' opener we all use today.

Tea Bag

Loose green tea was popular in China prior to the 1700s, but the demand for black tea overtook green tea in the later part of the century and soon milk was being added to create what would become Britain's national hot drink, the cuppa. By the 19th century, tea-drinking had spread, and Indian tea was being grown on a huge level, both to meet local demand and for exportation around the world. Everyone has an opinion on how to make the perfect cup of tea, and this great tradition perhaps finds its roots in the introduction of a small strainer known as a 'tea ball', which was temporarily placed in the water to prevent leaves being left loose in the pot. It seems logical then that the idea of tea bags may have come out of this. In fact, like many of the best inventions, the tea bag's arrival on the tea scene was an accident. In the early 19th century, Thomas Sullivan, a New York entrepreneur and merchant of fine teas, attempted to improve his marketing strategy by mailing samples of tea in silk sachets to potential customers. Sullivan's customers perceived the neat little bags as being intended for the brewing of the tea. Impressed by the 'innovation', they sent Sullivan letters commenting on his innovation, adding that the only problem with the bags was that the silk was too fine a material to allow the flavour of the tea to fully seep through into the water. Sullivan began improving the design, but it was Joseph Krieger who became known for delivering the first finished tea bag to the world. The innovation was raved about for its convenience and clean practicality. In the early days it only took off in professional catering circles, but after the Second World War the tea bag found its rightful places in homes and mugs everywhere. Many tea purists were opposed to this new-fangled method of brewing it, but the 1950s craze for luxury, must-have household gimmicks and gadgets of convenience ensured the tea bag was heartily embraced by those with no time to worry about fussy old traditions.

Thermos Flask

The origin of the Thermos flask starts with the vacuum flask, created in 1892 by Scottish scientist James Dewar. Dewar was experimenting with ways to store two of his many earlier discoveries, liquid hydrogen and oxygen, at very low temperatures. He knew that vacuums, unlike normally sealed vessels (where contained heat flows outwards comparatively quickly and escapes into the atmosphere), possess a very low pressure of gas, which inhibits heat flow and can therefore hold a near-constant temperature for extended periods. With this in mind, he created a flask with a vessel contained inside a slightly larger vessel of the same shape with a vacuum between the two. The inside flask was coated with silver to radiate the temperature of the stored liquid back into itself. In 1904, Reinhold Burger had the bright idea of applying the same theory to hot drinks. He had the idea patented and changed the name to Thermos flask, *thermos* being Greek for 'hot'.

Widget

The widget achieves the head you get on a draught pint down the pub, but out of a can in the comfort of your home – a device that brewers spent years trying to make. Guinness was the first company to get there. They spent £5 million in research over nearly five years and eventually patented their product in 1985. Invented by Alan Forage and William Byrne, the ICS (In-can System) is a capsule containing nitrogen. The capsule sits within the pressurised can, and the nitrogen is kept in the capsule by the pressure of the carbon monoxide in the beer. Pressure decreases when the can is opened, allowing the release of nitrogen, which, in combination with the carbon monoxide, gives the beer that all-important creamy head.

Wine

Evidence from a residue found while excavating an Iranian neolithic site showed that the fermentation process had already been honed for wine production by 5000 BC. From then on the basic principles of wine production remained pretty much unchanged. Over time wine became important in Egyptian, Greek and Roman cultures, and of course to Christianity, with improvements in wine-making technology and taste evolving very gradually.

Whisky

For many centuries, whisky has been distilled from cereal grains, with a range of grains including barley, rye, wheat and maize being used for different varieties. However, it is uncertain when whisky was first made. A well-known story is that St Patrick was responsible for introducing the process to England around AD 400, but this cannot be verified. Some historians believe that the distillation process for making spirits has been around since AD 800, and that it was discovered in the Middle East, eventually coming to British shores with Christian monks.

What we do know is that whisky has been produced in Ireland and Scotland for centuries. As one might expect, there is a debate about which of the two can rightfully say they were the first to make it! Some say Irish missionaries took it to Scotland, others say the Scots developed it themselves using distillation as taught by those Christian monks. Whisky first starts to show up in Irish and Scottish texts from the 1400s, and even today the spelling of the word depends upon where it comes from. 'Whisky' tends to be used for whiskies from Scotland, Wales and Canada, while 'whiskey' is used for the spirits distilled in Ireland.

CHAPTER FIVE:
NAUGHTY BUT NICE

'You bastards !...this isn't the stalls.'

Ice Cream

NAUGHTY BUT NICE

Bubble Gum

In 1906, Frank Fleer of Fleer's Chewing Gum Company was aiming to make a gum that could form bubbles when chewed and then blown through. He produced a gum that made rather poor bubbles and he called it Blibber Blubber. The drawbacks of Blibber Blubber were that it lacked flavour and was virtually inelastic. A more successful prototype, Dubble Bubble, is associated with the accountant Walt Diemer, Fleer's colleague. Diemer improved the existing gum to make it stretchy and pink. It was so coloured because there was no other food colouring in the factory. The stretchiness was, some say, due to added latex, but he claimed he didn't know how he achieved the desired effect. Wider rumours suggest Diemer didn't actually leave accountancy to create the gum, and only posed as the inventor when he stumbled upon the recipe formulated by someone else. Whatever the truth of the matter is, in 1928 Diemer got Fleer to sell it under the name Dubble Bubble and he hired some lively sales reps, coaching them in the novel art of gum trickery so that they could put on a decent bubble-blowing show for potential buyers all over America. The product became so commonplace that it was even incorporated into the Second World War rationing system. After the war more wacky flavours such as grape were made, and Fleer even televised a national bubble-blowing contest.

Chewing Gum

Chewing gum in one form or another has been around since ancient times. The Ancient Greeks were fond of masticating on mastic gum from the mastic tree, and Native Americans enjoyed getting their teeth around the resin found in spruce trees, a habit that was adopted by some of the first European settlers to New England.

Chewing gum as we know it was the brainchild of Thomas Adams, a New Yorker whose meeting with an exiled ex-President of Mexico, one Antonio Lopez de Santa Anna, led to him learning of chicle gum, a popular substance found in sapodilla trees that Central Americans love to chew. Adams tried and failed to market chicle as a rubber for tyres and other products. Chewing on some leftover stock one day, he realised that those Central Americans chewed chicle for a reason – it was pretty tasty! Shortly afterwards, he boiled some up, added some flavourings and began to sell it in 1869. Undeterred by criticism, he made a huge success of his business, and in 20 years his operation was a 250-man-strong chewing-gum empire!

Chocolate

The ultimate comfort food has been around since the time of the Mayans in Guatemala and Mexico around AD 500. The Aztecs harnessed the stimulating power of the cacao bean to make a reddish drink named *xocoatl*. The word chocolate derives from this, itself a combination of the words *xocolli*, meaning 'bitter', and *atl*, meaning 'water'. Its consumption was believed to increase fertility and, according to the 1519 conqueror of Mexico, Hernando Cortés, Emperor Montezuma consumed the drink in huge quantities. The recipe was then taken back to Spain, where a sweetened version proved very popular.

For many centuries chocolate was consumed as a drink, and the precise origins of solid chocolate are unclear. The first-known British reference to chocolate for eating is an 1826 advert for Fry's Chocolate Lozenges, marketing them as a food substitute for travellers. Fry's began making chocolate solely as confectionery in 1847 – their Chocolate Cream Stick was an instant hit. John Cadbury followed in 1849, selling his bars in Birmingham.

Condensed Milk

While crossing the Atlantic in 1851, New Yorker and inventor Gail Borden was deeply disturbed to witness a number of children dying from the consumption of contaminated milk. Having previously come up with some rather eccentric ideas, including a dry meat biscuit, he resolved to find a method of producing milk that wouldn't kill those who fancied a glass, and, recalling the method he had observed Shakers (members of the Society of Believers in Christ's Second Appearing) use to purify fruit juice, thought he'd give it a try on the white stuff. The technique involved heating vacuum pans, and Bowden made the mistake of thinking that removal of water was what made the milk safe, when in fact the heating process merely killed bacteria, thus sterilising it. This error is what made him call it condensed milk. One mistake he didn't make was not patenting the idea. He set up a factory in the 1850s, just in time for the Civil War when it so happened that Borden's milk became part of essential rations for the Federal Army. Its energy content and nutritional value was incredibly beneficial, and when the war was over soldiers returned home singing its praises – it soon became a hugely popular part of the American diet.

Croissant

What brunch or long breakfast would be complete without the beloved croissant? Well, the English breakfast, for one, as croissants taste awful with fried eggs and ketchup. Anyway, the crescent-shaped, freshly baked sweet pastry is said to have been made to mark the Polish victory over the Muslims in the Battle of Tours in 732. The shape is said to represent the crescent seen on the Turkish flag.

Doughnut

Doughnuts are lumps of sweet dough deep fried in fat, so, logically, they were first known as oily cakes – or *olykoeks* – in Holland where they started out. They were later introduced to the USA by Dutch immigrants. The claim is that, in 1847, Captain Hanson Gregory, the son of a baker, set sail with a hearty supply of his mother's treats. Halfway through the journey the crew came up against an aggressive storm that threw them around the deck. Captain Gregory was at the helm. In order to have both hands free to control the vessel, he impaled a doughnut on to a spoke of the ship's wheel, piercing it with a large hole through the centre. He took the story back to his mother and she loved it, especially because it gave her an idea for an improvement to her doughnuts. She had always experienced problems in achieving an evenly fried doughnut and often found their middles undercooked. She began punching a hole dead in the centre of the doughnuts before cooking, and found that they not only cooked faster, but also that they became more popular with the locals as they were easy to hold.

Ice Cream

Long before there was a Mr Whippy Ice Cream van round the corner selling '99' Flakes, and even longer before Ben and Jerry's started selling their incredible array of wild flavours, the ancients were in some way anticipating the arrival of the Cornetto. Roman documents report that in 62 BC the Emperor Nero (famed for his hedonistic and extremely decadent extravagances) ordered hundreds of slaves to travel long distances into the Apennine mountain range to bring back snow to be flavoured with fruit, honey and nuts and eaten as a sweet delicacy. Sorbet-like ice creams were made in Persia in 400 BC using rosewater and ice collected from mountain summits, and the Chinese concocted frozen desserts way back in seventh century. They experimented with a range of natural flavours, and their chilling process involved mixing snow with potassium nitrate to achieve temperatures below freezing. The earliest documentation of ice cream being enjoyed by a nation was found in Italy, who still produce the best *gelato* in the world. The arrival of ice cream as we know it today came about when one Monsieur Le Caveau introduced cream into the iced desserts in his Paris coffee shop. He became famous for it in the coffee-house world, particularly when he began having ice creams sculpted into various shapes for his favourite customer, the Duke of Chartres.

In the 1850s, milkman Jacob Fussell from

Baltimore realised he could use up excess milk from his rounds to make ice cream. He became the first mass-producer of it, which brought the price down and made it affordable to all. Cheap ice cream was available throughout America and then came to England. Prices were pushed down further by a monopoly created by the arrival of a bunch of Italian immigrants into Britain who became known as the Hokey Pokey Men, 'hokey pokey' being a corrupted version of the Italian for 'try a little'. A waffle-seller named E A Hamwi took the treat a step further in 1904 by inventing the ice-cream cone. Hamwi was working on his sugar waffle stall at an exhibition one slow afternoon when, a bit bored, he rolled up one of his waffles into the shape of a cone, and ran over to the ice-cream vendor in the adjacent booth. They filled the cone with it, and in that moment laid the foundations for the development of the '99'!

Jelly Baby

Everyone loves a jelly baby, and everyone has a favourite colour! One might say the jelly baby is a national institution, so it is fitting that they were invented for symbolic reasons. They were first created to symbolise peace at the end of the First

World War, and, apart from production being temporarily halted during the very un-peaceful Second World War because of rationing, they've remained a hit ever since.

Liquorice Allsorts

In the world of inventions, the all-time favourite Bassett's Liquorice Allsorts is a great example of an accident leading to something good. According to Bassett's, the colourful confectionery was created in 1899 when a salesman of theirs, Charlie Thompson, went to present the Bassett's sweet portfolio to a wholesale director. Thompson was having little success until he catastrophically knocked over all his trays of sweets. Within the sugary chaos he had created were nuggets of plain black liquorice mixed in with a range of novelty coloured shapes crafted from cream paste. This was the point at which the wholesale director became interested – he loved the contrast of all the shapes and colours and immediately placed an order with Bassett's. Liquorice Allsorts soon became a massive hit. The finishing touch to a confectionery classic was added when a Liquorice Allsort was shaped to look like the Bertie Bassett on the packet.

M&Ms

The phenomenon of M&Ms and the marketing sound bite 'The milk chocolate that melts in your mouth, not in your hand' were created from an idea that sprung from the Spanish Civil War. Forest Mars, who would go on to found the chocolate candy empire, observed that the sugar hit of choice for soldiers was chocolate drops. Noticing that they didn't melt in the scorching Spanish sun, he found that the key to their solid state lay in the hard sugar coating. Mars copied this and ordered loads to be made, naming them M&Ms, and he put them on sale in 1941. Thanks to the brand characters featured on the packet and in the advertising, as well as the packaging and the colour of the sugar shells, M&Ms quickly grew in popularity.

Wine Gums

Who, as a child, didn't wonder why wine gums are called wine gums? After all, they're not made of wine, and they don't taste of wine. The truth is nobody is quite sure why Charles Gordon Maynard, son and employee of Libyan sweet-shop owner Charles Riley Maynard, decided on that name for the sweets he created. What we do know is that when he presented the wine gums to his father in

1909 he nearly got the sack, for Daddy was a strict Methodist and teetotaller. But Charles convinced him that his recipe did not contain a drop of wine, and production went ahead. It has been suggested that the idea for the name may have come from its creator's perception that the colours of the gums in some way related to different types of wine. Beyond question, though, is the success of the product. In 1990, Trebor Bassett took over Maynards, and the sales figures for Maynards wine gums reached £40 million for 2002. Children's writer Roald Dahl was a massive fan, and he kept a jar of them by his bed so he could enjoy a couple of sly ones each night before he nodded off. Perhaps they offered some kind of inspiration for the creation of one of his greatest characters, Willy Wonka.

CHAPTER SIX:
CLOTHING AND APPAREL

Umbrella

CLOTHING AND APPAREL

Bikini

Early Greek paintings on jugs depict two-piece swimsuits worn by athletes as early as 1400 BC, but the somewhat sexier and skimpier bikini exploded on to the fashion scene with a bang in 1946. At this time, the Americans were carrying out tests with nuclear bombs on the island of Bikini in the Pacific Ocean. The two-piece is said to be named after the island because of the explosive effect it was likely to have on men. A year before the bikini's release, a very small swimsuit had featured in a Paris magazine, but perhaps due to common disapproval had not caught on. Louis Reard was the engineer who worked on the design of the bikini alongside the fashion designer Jacques Heim. Reard persuaded the model Micheline Bernardini to wear it and strut her stuff in front of the camera. The sheer volume of adoring letters he received in reaction to the photos proved that some members of the public were more than ready for such daring attire. After the predicable outrage from some quarters died down, bikinis were soon the must-have item for any young lady on the beach.

Bra

Bra is short for brassiere, an archaic French word for arm protector. The history of the bra is a lengthy one that begins in 2500 BC in Minoan Crete. At the time, a basic piece of material was strapped around the chest to support the breasts and keep them in place. It has been suggested that this may have been an early form of sports bra. As far as the development of the modern brassiere goes, the 16th century saw the full corset come into use, a design that tightly bound the breast and the whole of the torso to suit the fashions of the time. In 1893, American socialite Mary Jacob was dressing for an evening function when she became vexed with her whalebone corset. It looked rather daft under the tight silk party dress she had chosen, so Mary discarded it and instead tied two hankies together with ribbon, and used this to keep her vital parts where she wanted. She was so pleased with the result that she patented the design and began selling her 'Caresse Crosby' bras with limited success. Eventually, she sold the design rights and Warner Brothers Corset Company went on to make a vast fortune from Mary's idea. By the 1920s, it was in wide circulation and the corset became obsolete, only to be revived for costume wear, S&M and gothic trends years later. Today the bra is multi-functional: it can provide extra support, flatten, push down, volumise and, of course, push up!

Cardigan

Who would have thought that the invention of the humble cardigan, a garment as beloved by grandmas as it is by trendy young men, was inspired by war? Although a pretty useless soldier, James Thomas Brudenell, the 7th Earl of Cardigan, loved a touch of knitting. He came up with the cardie to keep his fellow troops cosy during the Crimean War, for it was bitterly cold. All very well, but maybe if he hadn't got his knitting needles out so often he wouldn't have made such a strategic mess of the Charge of the Light Brigade in 1854.

Jeans

In 1850, a man named Levi Strauss emigrated from Bavaria to San Francisco to take advantage of the gold rush. A forward thinker, he decided to carry with him large rolls of material with the intention of making tents and vehicle covers in the hope that there would be the demand to make it worth his while. One afternoon, a miner with whom Strauss had become friendly suggested that he started doing a sideline in workwear. The miner explained that he was fed up of wearing trousers that wore out so quickly due to the physical intensity of his job. Accordingly, Strauss designed the first jeans. They soon became popular in factories and down mines.

Shortly after, a design feature was introduced that remains on all the jeans we wear today – copper studs. There were attached to the stress points around pockets for further enforcement. The story goes that these copper rivets came about because of a young fellow who had a penchant for collecting rock samples, storing them in his pockets when out pursuing his interest. To help his jeans cope with the strain his precious rocks placed on them, he recruited a blacksmith to rivet his trousers. In the 1870s blue jeans complete with the rivets were patented. They remained popular for many years as sturdy work trousers. Soon after the war they began to be worn by the general public as leisure wear. Jeans became even more popular when the dark-blue material was tempered by the stone-washing technique. Films, especially Westerns, helped the jeans industry grow and grow. In the late 1980s, an infamous Levis advert featuring a man removing his jeans and washing them in a Laundromat had such an explosive impact on the popularity of jeans that the company had to temporarily pull the plug on the ad as production levels couldn't keep up with demand. These days, jeans are more popular than ever, with hundreds of brands offering an infinite variety of pre-torn, pre-stained and seemingly pre-ruined pairs available in clothes shops everywhere.

Miniskirt

Surely one of the enduring emblems of 'swinging London' in the sixties, the decade-defining miniskirt was the outrageous fashion item that caused a sensation on the international fashion scene. They were first seen in 1964 as part of designer André Courrèges's collection, and they hit Britain in 1965 when Mary Quant, an über-cool London designer, began selling skirts that were 6in (15cm) short of the knee.

Nylon

Nylon is a completely synthetic polymer, a large molecule made up of many repeated units of the same molecule, and is easily mass-produced. It was designed by Wallace Carothers, a polymer specialist working for DuPont in 1934 who was asked to synthesise a substitute for silk to compensate for Japan's decision to prevent exportation of it. Carothers tried various reactions until finally, in 1938, DuPont launched Nylon for use as toothbrush bristles. In 1939 they marketed it for use in what was to be their most successful product: ladies' stockings. Two years later they had sold nearly 65 million pairs. Interestingly, the word nylon has no particular meaning: apparently the prefix 'nyl' is arbitrary. It seems that the ending may have simply been added to be in keeping with 'cotton' or 'rayon'.

Safety Pin

The invention of the safety pin, or 'dress pin', came about as a result of one man owing another $15. American Walter Hunt was an inventor whose creations included a repeating rifle and a nail-making machine among a multitude of other gadgets, but he had failed to profit from any of them. One day in

1849, while ruminating about a $15 debt to his draughtsman, Walter nervously resolved to fashion an invention from a piece of wire he was distracting himself with and, sure enough, by the end of the day the safety pin was born! Walter's wife was especially pleased with the fruits of his labour, for she had often complained to him of pricking her hands with pins when trying to bend them to suit her needs. Walter immediately applied for a patent, writing, 'The distinguishing features of this invention consist in the construction of a pin made of one piece of wire or metal combining a spring or clasp or catch, in which catch the point of said pin is forced and by its own spring securely retained.' On receiving the patent he sold it for a mere $400 dollars to his creditors, who went on to make a fortune. And, yes, he cleared his $15 debt. A case of financial, as well as practical, necessity being the mother of invention.

Sewing Machine

In 1790, English cabinet maker Thomas Saint received a patent for his sewing machine, the design of which, although it did not work when built, almost prophetically anticipated many of the features of the mass-produced machines that would eventually appear 50 years later in the form of the lock-stitch models patented by Isaac Singer and Elias Howe.

In the meantime, the first chain-stitch machine was invented in 1829 by Frenchman Barthélemy Thimonnier, an impoverished tailor. The chain stitch (a series of loop stitches that forms a chain) was employed by embroiderers, and in this regard the Thimonnier represented a threat to those tailors who plied their trade without machines. So, when, in 1831, Thimonnier fulfilled an order for 80 machines from a Paris clothing manufacturer, it was a short-lived business triumph because a group of incensed tailors vandalised them all! With only one surviving machine, Thimonnier survived by selling small wooden machines until fortune looked as if it was about to shine on him again in 1845 when a manufacturer began to mass-produce his most up-to-date machine, capable of 200 stitches per minute. Things were looking rosy for all concerned until, that is, the 1848 revolution put an end to this fledgling industry, and Thimonnier eventually died in poverty.

In the meantime, in America, Elias Howe came up with the first lock-stitch machine in 1846. Lock stitching was very useful in that the whole chain does not come undone if one stitch is broken. Unable to get it manufactured there, he sold the rights to a producer in Britain, although eventually they fell out and Elias ended up back in America, and penniless, by 1849, only to discover that Isaac Singer had copied his idea and, what's more, was making a huge commercial success out of it. Howe sued Singer and ended up with his share of the profits from the Singer machines that became ubiquitous throughout the world over the following decades.

Shoe

Prior to 1500 BC, variations on the sandal were worn, though it is reported to be the Mesopotamians who were the first to join up the straps, covering the foot with material to provide warmth and protection. These early shoes were made out of leather and tied at the top with straps, also made of leather. About that time in Minoa, boots were made out of calf hide, and gave added warmth because they went further up the leg. It has been suggested that the Egyptians were first to shape the base of shoes according to which foot they were meant for, but rubber-heeled shoes, the first shoes designed to absorb shock, were not seen until 1899 in America when Humphrey O'Sullivan patented his design for them.

Trainer

Although 'sneakers' and plimsolls developed throughout the 20th century, the modern trainer started life in 1949 when Adolf 'Adi' Dasler created the first trainers with his company Adidas. Worn by several Olympic gold medallists, the trademark three stripes on the side of the shoe were designed to give extra support. The Puma brand was created by Adi's brother Rudolf. Trainer design has evolved constantly since then, with developments often focusing on trainer soles. American athlete Bill Bowerman famously filled a waffle iron with molten rubber in 1971. The resulting grid-like sole became a common feature of many training shoes. Bowerman went on to form Nike in 1979, and who hasn't heard of the air sole? These days, 'sneakers' – as they were originally called in America because the simple rubber sole allowed the wearer to quietly sneak up on people – are now a very hi-tech business, often with price tags to match.

Trousers

A type of trouser was worn in 300 BC in China, but only by soldiers. Descriptions of trousers in Europe appear from the 16th century. The word 'trousers' derives from the Gaelic *trouzes*. The word is plural because originally they were composed of two individual stockings, one for each leg, and attached to underwear or a jacket with clips at the top. Eventually the separate legs were joined, and by the early 17th century the 'fly' had been added.

Umbrella

Umbrellas have been around in one form or another since around 1500 BC, and were originally used as protection against the sun. The collapsible design was invented by the Chinese around AD 300 (they were also the first to add lacquer for protection against rain) and was adopted by the French and English in the 1600s. Louis XIII of France is recorded as having 'umbrellas of oiled cloth'. But, for nearly two centuries, umbrella use was strictly for the ladies. Men who chose to carry an umbrella were seen as effeminate, as superbly illustrated by records of Jonas Hanway, a philanthropist given credit for being the first man to carry an umbrella around London. Hanway suffered innumerable insults from coachmen, who perceived the protection from rain offered by umbrellas as a threat to their trade. Pious passers-by also commented, because they felt that Hanway was defying God by protecting himself from rain that the Almighty intended to fall on his head!

The steel-ribbed umbrella was invented in the 1850s but, as well as being expensive, many Victorian models were constructed out of wood and oiled silk, which made them hard to fold. Since then, multiple design improvements, with materials such as cotton, plastic and nylon often being used instead of silk, have made the umbrella a cheap and acceptable way of staying dry for both sexes …

Wellington Boot

Wellies came about during the Napoleonic Wars. Arthur Wellesley, the first Duke of Wellington, was dissatisfied with the performance of his army boots and asked his shoemaker, Hoby of St James's Street, London, to improve on the hessian boots that were standard issue. The improved design pleased him greatly with its soft calf skin, low-cut heels and mid-calf cut-off, and the duke can be seen wearing his tasselled wellies in an 1815 portrait by James Lonsdale. The boots went on to become essential kit for British aristocrats from then on. The rubber version that we are used to came along in 1852, when American entrepreneur Hiram Hutchinson got chatting to Charles Goodyear, whose recent invention of the rubber vulcanisation process he purchased for use in footwear. Before the year was out, Hutchinson was producing rubber boots and selling them in rural France.

In Britain, Wellington boots aren't just made for walking. The obscure sport of wellie wanging, which involves throwing Wellington boots as far as possible, has a number of ardent followers. The origins of the sport are unclear, but historians are fairly sure it has nothing to do with the Napoleonic Wars!

Velcro

Velcro is a fastening device that imitates nature. In 1941, engineer Georges de Mestral was walking with his dog in the Alps and became increasingly irritated by husks of burdock seeds sticking to his clothes. When trying to remove them, he was amazed at just how sticky they were. On closer inspection, he observed that tiny hooks on the husks were adhering to loops on the fabric of his clothes. He attempted to emulate this effect using synthetic materials to create strips with hooks and loops that would lock when placed together, eventually finding that nylon and polyester were the most appropriate for this. He named his invention Velcro. The 'vel' part comes from the French word for velvet (*velours*) and the 'cro' from the French for hook (*crochet*).

Zipper

Doing a lady's boots up, or making sure your trousers didn't fall down, was a time-consuming affair before the invention of the zip. All manner of arduous devices, like button-hooked laces and snap fasteners, sufficed to keep people's clothes on, but American Whitcomb L Judson, a man with little patience for the labour involved in dressing himself, tried to put an end to it all with his 'clasp locker or unlocker for shoes' in 1891, which was undoubtedly the precursor to the zip as we know it today. His invention caught the eye of Colonel Lewis Walker at the 1893 Chicago Exhibition, who began its manufacture with the Automatic Hook and Eye Co. The only problem was that his contraption didn't work all that well – his original zip kept coming undone! Needless to say the public weren't too enamoured by this, and the zip was largely ignored. It wasn't until Gideon Sundbank, an engineer from Sweden, came along in 1913 that an improved design was arrived at. Colonel Walker proceeded to market the Talon Slide Fastener but it was not until the Americans joined the war effort in 1917 that the zipper really achieved success. Perfect for the needs of the military, the demand for them was huge and Walker et al achieved runaway success. Once back from the war, soldiers spread the zipper word and the device was soon used for a multitude of civilian needs, mainly in foot and sports wear. The zipper phenomenon arrived in Britain in 1919. The Brits were initially wary, but soon accepted it for the marvellous device it is – although it was not until 1935 that zippers were used to keep men's flies up!

CHAPTER SEVEN:
IN THE BATHROOM

'I leave my two front teeth to my wife and the remainder to be divided equally amongst my rightful heirs.'

False Teeth

IN THE BATHROOM

Contact Lenses

Leonardo da Vinci had the foresight to come up with the idea of contact lenses. His 1508 sketch shows a flat lens at the end of a tube. Water was to be poured into the lens to create a curve through which sight could be corrected. But it was 1887 when the first workable contact lens was made by a German physiologist Adolf Eugen Fick. Made of brown glass, it was initially tested on rabbits, then on himself. But the lenses were wide – nearly 20mm in diameter – so they could only be worn for short periods of time due to the discomfort factor. Plastic contact lenses were nowhere to be seen until the 1930s, when a plastic/glass combination lens was introduced, and soft lenses popped up (and out) three decades later thanks to Czech chemist Otto Wichterle.

Cotton Buds

Ears were waxier before the cotton bud came along. Invented by Leo Gertstenzang in 1925, the idea for them struck him when he observed his wife cleaning their baby's ear using a toothpick with cotton wrapped around one end of it. His Infant Novelty Company began production using a machine to produce buds 'untouched by human hands'. They were first branded as Baby Gays, and were made from wood until the lollipop was invented in 1958 – thereafter paper sticks were used.

Dental Floss

According to statistics, only one in ten Americans floss daily, and half of them don't floss at all. The rest seem to floss now and again, but only when absolutely necessary. The main reason for such slack habits is probably the fact that flossing is mind-numbingly boring. The history of dental floss is equally dull: originally made out of silk, it was heavily endorsed by a New Orleans dentist in 1800. Jaw-dropping. In 1898, the Johnson & Johnson

Corporation received the first patent for dental floss. Oh, and nylon replaced silk in the 1940s because it worked better. Yawn, and open wide, perhaps say 'aahhh'.

Disposable Nappy

It is perhaps unsurprising that the idea for a disposable nappy was born out of a mother's concern. As a mother of two faced with the problems of nappy-leakage, nappy-washing and, above all, nappy rash created by traditional terry towelling covered in rubber pants, Marion Donovan was convinced there must be an easier way to deal with the issue of infant toileting. Coming from a family of inventors, it was in her blood to find a solution.

One day she began experimenting by cutting up her plastic shower curtain to use as an outer layer for a cotton nappy, and found it kept her baby's bed dry. While trying out other fabrics, she eventually landed on parachute nylon and decided this worked best. She began selling her breathable, non-leak nappy cover in 1949. The next step was to create the disposable paper nappy to be used in conjunction with the cover. Her product was an instant hit with mothers, so much so that she could not make them quick enough to meet the constant demand, so Marion sold the patents to her creations for $1 million in 1951. Proctor and Gamble eventually produced the resulting commercial product a decade later, and a truly global brand was born. That's right: Pampers.

False Teeth

The first false teeth were a far cry from the dentures seen in Granddad's glass by the bed. Made by the Etruscans around 700 BC, they consisted of dental plates combined with gold frames that held teeth made from bone, teeth from other humans, and even teeth from animals. Over the centuries that followed, other materials were tried with varying degrees of success, but during the 18th century ivory became the material of choice for replacement teeth. The only problem was that the ivory tasted revolting and soon rotted away. Teeth from the deceased became popular, and hippo teeth were also experimented with, but the problem of rotting persisted. A solution was needed, and in 1770 the use of porcelain was introduced by French apothecary Alexis Duchateau. After many teething problems, he eventually cracked the longstanding puzzle of sizing the porcelain correctly so that it could be fired in a kiln, and made himself a set of choppers that he wore until his dying day. His porcelain dentures were soon being manufactured, wiping the toothless grin from many a wearer's face.

Hairdryer

The first handheld hairdryer came about in 1920, produced by the Hamilton Beach Manufacturing Company in America. Up until then, there had been no motor small enough to power a fan to blow air over an electric heater – a fan and a heater being the essential ingredients of any hairdryer worth its salt! Miniature heaters using high-resistance wires, and based on the principle of the light-bulb filament, were already in existence, so all it took was the brains of American Chester Beach who came up with the first small electric motor in 1905. Beach's ideas, combined with the business brawn of Fred Osius and LH Hamilton, eventually produced the metal-cased Cyclone in 1920. Plastic hairdryers came about in the 1950s, and various permutations of the basic design eventually led to the market we have today, with a dazzling array of fancy hair-drying equipment available in a high street near you.

Ever wondered why using a dryer lets you control and shape your hair so effectively? If so, here's a little science. The heat accelerates the formation of hydrogen bonds within the hair, and these strong bonds give a better 'hold' to whatever style you choose, albeit a temporary one.

Lipstick

Traces of finely crushed semi-precious stones suggest this was the way the Ancient Mesopotamians concocted their lipstick mixes over 5,000 years ago. Fucus, an algae plant matter, was used by the Ancient Egyptians for the purple and reddish colour, and mixed with iodine for a brown base. Bromine mannite was also added – until it was discovered to be deadly. In Cleopatra's time (around 50 BC) crushed cochineal beetles were a source of red pigment used in early lipsticks, and some time after this a sparkly, shimmery look was achieved with the addition of silvery fish scales, one of the by-products of herrings in the fish trade. During Elizabeth I's reign, it was all the rage to wear fully glowing scarlet lips against a pale white face, and improvements were made to natural formulas by incorporating blends of plant extract and beeswax into lipstick formulas. Post-Second World War movies further emphasised the sexiness of luscious red lips. New wax-free, semi-permanent formulas caused sales to shoot up when Maurice Levy patented the modern twisting lipstick dispenser. The 'lippie' became the must-have item in every girl's handbag.

Mascara

Mascara, adapted from the Spanish *máscara* and the Italian *maschera* – both meaning 'mask' – was dreamed up by New York chemist TL Williams as a way of helping his sister Mabel look good enough to prevent her man from cheating on her. In 1913, he mixed Vaseline with coal dust and applied it to his sister's eyes. Apparently, it did the trick as Mabel soon married her fella! So pleased was he with the creation that Williams began production of Maybelline (a combination of his sister's name and Vaseline), and by 1917 it was a hit across America, with Maybelline soon becoming a major cosmetic producer.

Pregnancy Test

Finding out whether you're pregnant these days can be a costly business. Suppliers of the simple, reliable test can command a premium price – it's a must-have item of sorts, because, when you need to know, you need to know! In Ancient Egypt you wouldn't have needed to shell out much for a test. In fact, if you lived in the countryside, you probably wouldn't need to shell anything out at all – you could simply collect grain and urinate on it! A pile of wheat and barley seeds was mixed together in a 50:50 ratio before a woman stood over it and took a pee. If anything sprouted, the test was positive, and Egyptian records state it was more or less accurate. Furthermore, if the wheat germinated, it was predicted to be a girl and, if the barley germinated, a boy was on his way.

In the early 1900s, Bernhard Zondek developed the first modern pregnancy test. It worked by detecting the presence of the human chorionic gonadotropin (HcG) hormone in urine, and involved injecting frogs with the urine of the woman – if the woman was pregnant the frog would produce eggs within a day. The same principle is used today, but things are little more sophisticated. After 1977, rather than nipping into the toilet with a frog up her sleeve, a lady could buy a kit. This consisted of a neat, discreet pen that uses a specific antibody to detect the presence of the hormone in the urine. The result could be seen within a few minutes. Two blue lines for pregnant and one line for not pregnant. The same test is used all over the world today.

Razor

The custom of removing one's hair to look respectable has been around since Egyptian times – the Egyptians simply loved it and would remove every hair from their bodies using blades forged from gold or copper, as well as flint stones and shark's teeth. From the 1700s onwards, the cut-throat razor became the norm for gentleman's shaving until King Camp Gillette progressed things somewhat with the development of the disposable razor in the early 1900s. The idea had come from his boss, disposable-cork inventor William Painter, who suggested to Gillette that, if he invented 'something which will be used once and thrown away', he'd create a market where the customer would keep coming back for more, thus lining Mr Gillette's pockets. Thinking on this, Gillette observed what a tiny proportion of a cut-throat razor actually comes into contact with the skin. Realising that the razor's edge is the only significant part needed for cutting, he resolved to create a small sharp blade with little unnecessary backing. Undeterred by steel-producers who told him it wasn't possible, and with the help of William Nickerson (sole employee of Gillette's fledgling American Safety Razor Company), he began

143

producing safety razors in 1903. Initially, sales were dismal – less than 200 blades sold nationwide – but things picked up somewhat, with over 10 million razor blades sold by the end of the following year's production. Another boost for the company was the British government's decision to make Gillette razors standard issue to soldiers during the First World War.

The first successful electric razor was invented by ex-US Army Lieutenant Colonel Jacob Schick, who was so miffed at having to shave in freezing water while in Alaska that he was determined to come up with a dry alternative. His 1928 machine had an external motor to drive the shaving head, and had to be started with a turning wheel. After an initially frosty reception from the American market, and the impact of the Great Depression (when survival became more of a priority than a clean shave), sales improved and his razors achieved worldwide success.

Roll-on Antiperspirant

Antiperspirant appeared in the late 19th century, but development was slowed by the fact that sweating was a hush-hush subject. Everybody wanted a solution to the problem, but nobody wanted to discuss it. Mum was the first company to sell the early formula, which was essentially a paste containing zinc sold in a can. It was applied to the underarm using one's hands. By the early 1900s aluminium chloride was added into the mix. The combination of aluminium and zinc successfully prevented embarrassing and malodorous sweats by keeping the area dry, but nobody knew the exact mechanism by which it worked. The belief was that, somewhat unpleasantly, it was to do with the pores being blocked, thus blocking the flow of perspiration.

The problem of a neat way to apply the rather messy antiperspirant liquid remained. A contraption was needed that would enable application against the force of gravity (because the device would have to point upwards into the armpit). One day, the solution was chanced upon when one of Mum's directors was chewing on the end of his pen. Daydreaming away, he pondered the way in which ballpoint pens release ink and noted that this principle could well be copied for use in deodorant. By the mid-fifties, the roll-on had passed all the required tests and was sold both in the US and UK. Today, there are many products competing to be most effective at keeping people smell-free, including creams, aerosol sprays, liquid sprays and even deodorising wipes. But the roll-on remains the favourite for many.

Soap

There is much uncertainty about when soap was first used for the purpose of human hygiene, but what we do know is that from around 2800 BC the Babylonians were producing a soapy substance for one purpose or another. Clay cylinders were used to boil up fat and ashes, and the result was a soap of sorts. This kind of soap would not have made a great lather, and it is thought that early soap may have been used for some medicinal purpose until around AD 150 when the Romans began to use it as we do today. Pliny the Elder (7 BC–AD 53) was the first to write about the use of soap in his *Historia Naturalis*. He mentions it being used for dyeing hair.

A common, and interesting, myth surrounds the history of soap, and claims that the word came about from rain washing fat and ash created by Roman animal sacrifices down Mount Sapo into the river Tiber, where women would be washing clothes. When the substance mixed with clay, the soapy water that resulted helped them clean their clothes. However, nobody knows where Mount Sapo is, nor where the legend comes from, so it is hard to give it any credit, especially because the fat from animal sacrifices was eaten by humans, and the entrails left to the gods. A slightly less captivating explanation of the word's origin is that soap simply derives from the Latin word for it, *sapo*. However, the latin word may have come from the Mount Sapo, so perhaps the myth is true after all…

Suntan Lotion

Coco Chanel is credited with making suntans trendy in the 1930s. She came back from her holidays in the south of France with a suntan, and before she could say 'I'm peeling' half the women in the West were out baking themselves in the heat. This, of course, created a market for cream to prevent sunburn. First past the post was French chemist Eugene Schueller, who founded L'Oreal. He created the first preparation of Ambre Solaire in 1936, which at that time was an oil-based, perfumed liquid. But credit for the first truly effective suntan lotion product is usually given to Benjamin Greene, a pharmacist who responded to the needs of soldiers who were getting fried in the sun during the Second World War. The gloopy, red substance he prepared in his wife's oven worked a treat for keeping the rays at bay, and 'red vet pet' (red veterinary petroleum), a Vaseline-like substance, saved many a soldier from painful tan lines. However, once the war was over it never made it to the cosmetics counters in department stores – perhaps because it was thick and sticky. And red!

Tissues

Probably the most well-known brand of tissue is Kleenex, made by Kimberly-Clark. Made from cellulose wadding originally used as clinical dressings in the early 20th century, they were later adapted to be used as filters for gas masks during the First World War. After the war, the product didn't become obsolete as there was a huge surplus of unused tissues. Initially Kleenex tried to market them as make-up removers, but they were soon found to be equally useful for dealing with a runny nose caused by the common cold. Accordingly, they went on sale marketed as a cheap, disposable alternative to the cotton handkerchief.

Toothpaste

Most people would probably agree that toothpaste has a pleasant enough flavour and smell. Not surprising, then, that we use it to freshen our breath as well as to clean our teeth. However, we may not have enjoyed fourth-century Egyptian potions that were said to contain salt and pepper. Other more gruesome potions in the past have included insects, burned foodstuffs and even urine!

More effective cleaning powders began to be used for cleaning teeth. Abrasive substances such as crushed coal were tried, as was bicarbonate of soda, which both cleaned and whitened. The wet formulas, which became known as 'toothpaste', came about from the addition of water to powdered concoctions along with glycerine, which helped to thicken it. Eventually fluoride was added when it was discovered to have properties that prevent teeth from rotting.

In the mid-1890s Colgate, who were soap and wax merchants at the time, marketed toothpaste in a soft malleable tube. This squeezable packaging replaced toothpaste sold in jars.

Vaseline

In the late 1850s, Robert Cheesebrough was a chemist slightly down on his luck. Having been earning a living creating products from the oil in sperm whales, the flourishing petroleum market had pretty much cut short his career. However, his luck was about to change. In 1859, he paid a visit to Titusville, Pennsylvania, to investigate a recent oil strike and see if he couldn't adapt this oil into some new products. During his visit, some rig workers complained of a substance that built up around and clogged their oil drills. Intrigued by what had been nicknamed 'rod wax', and more intrigued still by the great claims made by the workers about its ability to speed up the healing of their cuts and grazes, he put his chemist's hat on. On investigation, he discovered the substance to be petroleum jelly. The wax was black, but Cheesebrough was able to distil a clearer, lighter product from it. Testing it on a number of self-inflicted minor cuts and burns, he found that the wax did indeed help them heal up more quickly. And so Robert found himself with a new product on his hands (and his cuts). He called it Vaseline after the German for water (*wasser*) and the Greek for oil (*elaion*), and travelled America giving himself many cuts along the way in order to convince people to buy it. And, of course, they did. The global success of his product made him a very rich man. Vaseline is today used for a host of purposes, and of course still helps to heal cuts, though it is merely the seal created by the substance that protects the wound, not the action of some wonder-chemical in the Vaseline as originally thought.

CHAPTER EIGHT:
MEDICINE AND HEALTH

'Well I didn't pack the sticking plasters either !'

Sticky Plaster

MEDICINE AND HEALTH

Antiseptic

Antiseptics protect by destroying micro-organisms. The word means 'against' or 'in opposition to' sepsis, which is putrefaction or rotting caused by invading pathogens (infectious agents). Early attempts at creating antiseptic environments were made by Ignaz Semmelweis during the 19th century. Semmelweis was a consultant who told medics in a Vienna hospital to clean their hands with disinfectant to avoid fatalities when delivering babies. His advice was perceived as strange and served only to annoy people. Rather than following his instructions, they largely ignored them. Eventually Semmelweis was dismissed for the bad atmosphere he had created!

In 1864 Louis Pasteur experimented with his earlier hypothesis of the existence of germs and how to control their activity. He experimented with milk because it was something he had seen go off very quickly (see page 93). When eventually published, his germ theory described the conditions and temperatures required to inhibit destructive organisms in milk. His work on the pasteurisation of milk brought germ theory to acceptance and, in 1865, led Joseph Lister to apply them to a clinical setting. Lister hoped that, if he could eradicate harmful bacteria in patients and their wounds, he would reduce the death rate both during and after surgery. This was especially important as it was around the time the anaesthetic was coming into use, and the number of operations was on the increase. Lister made everybody wear clean clothes and 'scrub up' before operations. He also insisted on spraying carbolic acid everywhere, including on to patients' wounds, so that he could kill every possible germ that may be lingering. A significant drop in surgery-related deaths was recorded after Lister's measures were in place, and for his efforts he was made President of the Royal Society in 1895 and Baron Lister in 1887. Joseph Baron Lister is still remembered as the founding father of antiseptic surgery. Nowadays, antiseptic is commonplace in the household for both minor wounds and cleaning floors and surfaces.

Artificial Respirator (Iron Lung)

It was American Philip Drinker who devised the first artificial respirator in 1927. Used to keep people breathing when breathing function is permanently suspended, the first model consisted of a vacuum cleaner that controlled the air pressure in a sealed box. By 1931 a more advanced version had been arrived at, and the first iron lungs were employed in medicine.

Aspirin

Aspirin finally came on to the market in the late 1890s, quite some time after it was first discovered as a treatment for pain. The medicinal component salicin was used as early as the fifth century BC in Hippocratic medicine for pain and fever, having been found in the inner bark of the willow tree.

Centuries later, in the early 1800s, the crystalline form of Salicin and converted into salicylic acid. Later it was discovered that salicylic acid could also be extracted from the meadowsweet herb. The acid had painkilling properties, but severe side effects such as internal bleeding prevented salicyclic acid from being safe as a painkiller for humans. It was fortunate that in 1853 Charles Gerhardt found neutralising with sodium salicylate and acetyl chloride would form the safe yet efficacious product acetosalycilic anhydride. Unfortunately, Gerhardt's lack of interest in getting the product to market prevented it becoming available to the general public. It wasn't until 1897, when Felix Hoffman and his colleague's interest in it was sparked, that the molecule was modified to form acetylsalicylic acid. Hoffman found acetylsalicylic acid so effective in treating his father's agonising rheumatoid arthritis that he took his findings further and persuaded his employers, Bayer, of the compound's analgesic qualities. Bayer scored the patent and marketed the drug under the name aspirin, which is derived from *Spiraea ulmaria*, Latin for meadowsweet. It was first marketed in powder form and wasn't sold as tablets until 1915. Interestingly, heroin was marketed as a painkiller alongside aspirin. Heroin was another of Hoffman's projects and was initially the more successful painkiller. However, as soon as its highly addictive nature was discovered, sales of heroin were overtaken by aspirin.

Breathalyser

Perhaps unsurprisingly, this device for detecting blood alcohol levels, or BAC (blood alcohol concentration), based on a breath sample, was invented by Dr Robert Borkenstein, a man of the law in 1954. Frustrated with carting drunkards off for a blood sample, the captain with the Indiana State Police (who later became a university professor) developed a device that ascertained BAC using chemical oxidation and photometry. The glass tube in a breathalyser contains a variety of chemicals that change from orange to green in reaction to alcohol. An individual breathes into a bag to produce a certain quantity of air, and this passes into the device. The extent to which the green passes along the tube determines whether or not the driver is over the limit.

Condom

While the poetic adage 'Hey you, don't be silly, Put a condom on your willy' is relatively recent, the desire to fool around without consequences is nothing new. As far back as 1300 BC, condom-type sheaths are depicted in Egyptian art, and paintings dating back to AD 200 depicting similar penile apparatus have been discovered in France. But it was Italian doctor Gabriel Fallopius who first wrote about his condom design, which was 8in long and fabricated from linen. The wearer held it on by tying it with ribbon. Most early condoms were used more as protection against disease than as contraceptive devices, and all were made of either animal gut or fish membrane. It was not until the vulcanisation process came about in 1839 that rubber became the material of choice. Latex condoms as we know them emerged in the 1930s.

Contraceptive Pill

With the goal of 'fewer but fully desired children', Austrian Ludwig Haberlandt of the University of Innsbruck set about researching hormones and their possible role in contraception in 1929. His discovery that they were effective in the prevention of pregnancy was developed further by other scientists, but the problem was that the active hormone was progesterone, which was very pricey to isolate and didn't work when swallowed. It wasn't until Dr Gregory Pincus came along and worked out that progesterone could be synthesised from Mexican yams that things really started to move along. Having formed the company Syntex, his colleague Carl Djerassi's research led to eventual synthetic success in the early 1950s. Initially the hormones were employed to alleviate menstrual problems, until Pincus was asked by American Margaret Sanger of the Planned Parenthood Movement to conduct research into an oral contraceptive. Margaret was a working-class Irish-American who aimed to 'change the destiny of mothers whose miseries were as vast as the sky'. She was after something 'harmless, entirely reliable, simple, practical, universally applicable and aesthetically satisfying to both husband and wife'. After five years of hard work, Pincus and John Rock developed a pill made out of progesterone and oestrogen. Following clinical tests in the 1950s, the first commercial pill became available in 1960. It was no coincidence that the sexual revolution came soon (too soon, some might say) afterwards.

Prozac
Antidepressant medication has been around for many years, but Prozac arrived with such a media explosion that many people perceived it as the first wonder drug for a disaffected generation. It was in fact the third antidepressant of its kind to be available to the general public. Prozac is an SSRI (selective serotonin reuptake inhibitor) and was discovered due to findings in the 1960s that

serotonin (5-hydroxytryptamine, or 5-HT) is involved in the biochemical mechanism contributing to depression. Serotonin is a brain chemical that produces happiness and well-being in humans. It occurs naturally, and is produced and used in a complex process that involves it being removed to make way for fresh supplies. The part of the brain that deals with serotonin often uses the chemical too quickly in depressives, and the brain can't keep production levels up. Scientists realised that, if they could get the happy chemicals to hang around the right place for the right amount of time, they might be able to help people feel as pleased with things as they ought to be! So they looked for compounds that would reduce synaptic uptake of serotonin (the rate at which it is absorbed once produced) and alleviate depressive symptoms by making more serotonin available in the brain. Zimeldine was the first compound of this kind to be marketed in 1982, though its success was hindered by several side effects. Fluvoxamine, also an SSRI, was made commercially available in 1994, but it was prescribed for obsessive compulsive disorder. Ray Fuller was behind the invention of Prozac, chemically known as fluoxetine hydrochloride. He found the formula the most appropriate for the treatment of depression, and it possessed the least side-effects. Fuller won the Pharmaceutical Discoverer's Award from the Mental Health Research Association for his groundbreaking work in depression, and in 1974 pharmaceutical company Ely Lilly patented the compound Lilly 10140, renamed it Prozac for marketing purposes ('pro' having positive connotations, and 'zac' just sounding zingy, it is assumed) and it is now prescribed for over 45 million patients worldwide.

Spectacles

In ancient times semi-precious stones were crushed to make early lenses. The Roman emperor Nero is said to have enjoyed a superior view of gladiator fights when he started holding an emerald in front of his eyes. It is not quite known when the technology of concave and convex glass was discovered to assist vision. Reports often cite a Franciscan monk in Italy as the inventor of the first eyeglasses: these were held on the nose with a clip and it would be another few hundred years before the arms were added to hang off the ears. In the 18th century Benjamin Franklin became known for inventing the first bifocal lenses, alleviating problems for people with both long- and short-sightedness who had previously needed to keep swapping glasses for reading and vision.

Sticky Plaster

Johnson & Johnson employee Earle Dickinson had an extraordinarily clumsy wife named Josephine – accident-prone to the point that huge amounts of Earle's leisure time was spent dressing her continual supply of wounds with the adhesive tape and gauze produced by his company. Earle solved this by pre-preparing dressings for her to use. His method was simple yet ingenious: taking a length of surgical tape, he unrolled it and spaced small pieces of gauze on to it, before rolling it up again with crinoline on top to stop it sticking to itself. *Voilà* – whenever his wife needed to bandage another cut or graze, she had a ready supply of dressings!

Thinking his idea might come in handy to others, Earle presented the idea to the company, who felt it was a marvellous invention and began production at once. Johnson & Johnson's Band Aid was launched in 1921. First sold as a roll from which strips were cut, sales were initially slow, but things changed when pre-cut strips were marketed, and Band Aid became something no household was without.

Vaccine

The practice of administering vaccines, known as variolation, existed for many years in the Middle East before it was recorded in Europe. It was known that, if a small or weaker strain of a disease was administered to a person, they would become immunised against the full-blown strain. In 1796 Edward Jenner observed that those who had suffered from (or been exposed to) cowpox, a weaker strain of the deadly smallpox virus, seemed to be protected against the latter. A local dairymaid was in agreement, saying that she had come out in cowpox lesions on her hand from milking cattle yet she had not been struck with smallpox despite being exposed to it. Jenner took some diseased serum from an open wound on her hand and injected it into a small boy. He recovered perfectly from this, so Jenner injected him with a potentially fatal dose of smallpox. The boy was fine because his immune system had been adapted by the weak virus, enabling it to recognise and fight the stronger virus. Jenner didn't know this at the time, but at least his theory was correct and could be applied. He coined the term vaccination as *vaccinus* is Latin for of, or from, the cow. Jenner's refined vaccination techniques went into widespread practice, saving hundreds of lives.

X-ray

Thanks to the luck and persistence of Wilhelm Röntgen in 1895, the X-ray was introduced. Röntgen was a physicist in Germany who was experimenting with cathode rays and discharging them through tubes, within which were near-vacuum conditions. He wanted to look at the effects of discharging an electric current inside the tube. One day he suddenly noticed green light projected on to a screen on the other side of the room. He realised that there were dense objects between the tube and the screen, including solid wood, and wondered where on earth the light was coming from! Röntgen eventually concluded that there was only one possibility – the cathode rays passing through the tube were being discharged inside the tube and further rays were then emitted through both the glass and the objects and hitting the screen. He also found that it was only particular materials that picked up the rays and fluoresced. More surprising to Röntgen was that, when he raised his hand in front of the screen, he saw the outline of his bone structure. He then raised his wife's hand in front of the screen but this time he captured it on to photographic paper. This was the first detailed anatomical photograph and, on realising the implications for medical science, Röntgen excitedly set about publishing his findings. He couldn't think of a name in the early stages and temporarily called it radiation X. Against his wishes, the X remained and the term X-ray stuck to describe the type of ray emitted under these conditions.

CHAPTER NINE:
AT THE OFFICE

'O.K Jameson...two years on, we're older and wiser: let's stop these office pranks and shake on it.'

Superglue

AT THE OFFICE

Bubble Wrap

Bubble wrap was originally designed in the 1950s by Alfred Fielding and Marc Chavannes of the US company The Sealed Air Corporation. Their brief was to come up with a 'space-age' decoration for internal walls. Some say it was intended for use as a wacky wall hanging; other accounts describe its original purpose as wallpaper. Either way it didn't really catch on.

Fortunately for its inventors, it was found to be more useful as packaging to protect goods during transportation. The air-filled pockets were particularly good at absorbing shocks on bumpy roads and choppy seas. Additionally, the plastic in between the bubbles was designed to have anti-static properties to further protect delicate electronic parts.

A third 'use' for bubble wrap is that it's great for popping. A peculiarly pleasurable activity, its universal appeal is so widely acknowledged that there is even a website created for the addict to have a virtual bubble-wrap-popping session (www.virtual-bubblewrap.com). So, whenever you're short of a bit of air-filled plastic, just switch on your computer!

Correction Fluid

The first correction fluid was invented by a high-school drop-out who had no aspiration to become an inventor. Bette Nesmith Graham learned to type soon after leaving school, and some years later, while working as a secretary at a bank in Texas, she observed sign writers painting over mistakes and thought that a similar principle might work for covering over typing errors. Mixing some paint to match the colour of the company paper, she found that it worked, and other secretaries were soon borrowing the paint to use on their own work. Bette worked to make the fluid faster-drying and, after it was rejected by IBM, she set up on her own. In 1956 her Mistake Out Company was launched, and Bette and her husband Bob devoted themselves to it full-time. They were soon amassing a fortune, and in 1968 renamed the company Liquid Paper. Gillette eventually bought them out in 1999 for nearly $50 million. And no, that's not a typing error – $50 *million*.

Paper Clip

The beautiful simplicity of the paper clip makes it a magnificent piece of stationery. However, the story of its invention doesn't really involve much more than a few men having fiddled around with a piece of wire until one of them came up with the best design and patented it! That man was Norwegian Johan Vaaler, and the year he filed the patent was 1899. A few years later, English company Gem Manufacturing Ltd added the 'loop within a loop' feature, and that, as they say, was that.

However, what is perhaps a little more interesting about the paperclip is its place in the history of political protest. Because Norway had no patent laws, Vaaler's original patent was taken out in Germany. This turned out to be ironic because, during the Second World War, Norwegians wore paperclips in protest against Nazi occupation – having been forbidden to wear buttons bearing the Norwegian king's insignia, the Norwegians used the clips instead.

Photocopier

American Chester F Carlson was a physicist and
researcher for Bell Telephone Laboratories who
found himself out of a job when the Depression hit.
After eventually finding work in the patent
department of a New York electronics company, he
tired of having to copy endless technical drawings
and documents by hand using carbon paper and the
like. Nagging arthritis and short-sightedness made
an already monotonous job even more of a challenge
for Chester, so he set about designing a copier to
make things easier. Aware of Hungarian Paul
Selenyi's discovery that light has a bearing on the
electrical conductivity of a number of materials, he
set up experiments that tried to make a 'shadow', or
copy, of something using altered conductivity.
Within just a month, he had managed to copy '10-
22-38 Astoria' (the date and place of the experiment)
from a glass slide on to a charged, sulphur-coated
zinc plate, and then some waxed paper. Despite his
breakthrough in what he called electrophotography,
Carlson could not convince any companies that his
idea would be of use to people who already had
carbon paper and blueprinting as a means of copying
documents. IBM and General Electric weren't
interested, and it was not until 1947 that the

American Haloid Corporation bought the manufacturing rights and made the first Xerox machine in 1959. Why Xerox? Well, Carlson and Haloid coined the word 'xerograpy' from the Greek words *xeros* and *graphein*, which translate as 'dry writing'. The Xerox 914 sold so well that the Haloid Corporation became the Xerox Corporation in 1961. Carlson made millions from his invention but gave most of it away to charity before he died in 1968.

Post-it Note

The arrival of the Post-it Note is an example of combined strengths creating something so useful that many simply couldn't do without it. There is a 'doer' (Spencer Silver) and a 'thinker' (Arthur Fry) in this story, and they both worked at 3M (originally the Minnesota Mining and Manufacturing Company). In 1969 Silver was aiming to concoct a heavy-duty sticky glue when, to his surprise and disappointment, he found his glue wasn't very sticky at all. In fact, it was so unsticky that it didn't even leave a mark on paper or surfaces to which it loosely adhered. Strange! He thought long and hard but failed to come up with any use for it. His only idea was that it might be used as a backing for notice boards to obviate the need for drawing pins. Four years later, Arthur Fry, a regular chorister who used bits of paper to mark the places in his hymn book before choir, was frustrated by how often they would fall out and make a mess. Constantly losing his place in the book was too often causing him to miss the start of the song, which is any chorister's worst nightmare! He needed something that would temporarily mark his place (and *stay* in place) without defacing or tearing his hymn book in the process. Fry remembered his colleague Spencer's unsticky glue and set about covering small bits of paper with it. They worked a treat and didn't damage the pages either. The idea was presented to 3M, who began selling the Post-it Note worldwide. Along with other must-haves, like the coffee machine and the office clown, they are a quintessential part of any self-respecting workplace.

Rubber Band

In 1823 Briton Thomas Hancock's prototype for the rubber band was produced by a machine of his, the masticator, which 'chewed up' scrap rubber and made bands out of it. The problem with Hancock's masticator-made bands was that, in cold conditions, they became brittle and would snap under mild pressure. In order to make the rubber more durable, it needed to be treated in some way. Various efforts were made to do this but it was finally managed by US inventor Charles Goodyear, whose claim to fame is well known for the technique of the vulcanisation of rubber in 1839. Vulcanisation 'cures' rubber under high temperatures and relies on the addition of sulphur. The process strengthens the rubber while maintaining its elasticity, enabling it to remain intact under conditions of great stress. Goodyear said his findings were the result of extensive trial and error, but others claim they must have been serendipitous. Sadly for Goodyear, six years after he claims to have made his discovery, Englishman Stephen Perry was granted the patent for rubber bands. His company Messrs Perry & Co began selling them in London right away. They have since become indispensable, both for holding things together and for flicking at teachers.

Sellotape

Although a trademark itself, rather in the manner that the word Hoover is used to refer to all vacuum cleaners, Sellotape is used to refer to all clear adhesive tape. The need for such clear adhesive tape arose in the 1920s when flower and fruit sellers began wrapping their products in the new wonder-product of the time, cellophane – they needed a tasteful, clear sticky tape to seal and hold the cellophane in place. It was the inventor of paper masking tape, American engineer Richard Drew, who came up with the goods. He worked for the Minnesota Mining and Manufacturing Company, which later became 3M, who produced a range of adhesive products. He simply coated strips of cellophane with adhesive, and in 1930 it went on to the market as Scotch Tape. In 1937 the British equivalent, Sellotape, came on to the market courtesy of George Gray and Colin Kininmonth, who produced it in Acton, West London.

Superglue

As kids, we all heard the scare stories about Superglue – people stuck to toilet seats, lips sealed together forever, and much worse – and boy did they work at getting us to treat the stuff with extreme caution! Superglue has been around since 1958, and was invented by American researchers Harry Coover and Fred Joyner. While experimenting with the chemical cyanoacrylate, a potential substitute for spider silk that was used for the cross hairs in gun sights, they realised it was just way, *way* too sticky. But they also realised they might have a new wonder product on their hands. They were right, and famously demonstrated the sticky stuff's incredible power on the television programme *I've Got a Secret* by lifting host Garry Moore into the air by two steel plates held together with a single drop of Eastman 910 adhesive, the first name for the product. During the Vietnam War, cyanoacrylate sprays were developed and issued to soldiers due to the chemical's marvellous ability to stop bleeding and bind wounds.

CHAPTER TEN:
MONEY MAKES THE WORLD GO ROUND

'I like the idea chief, but we'd better start
with something simpler.'

Vending Machine

MONEY MAKES THE WORLD GO ROUND

Banks

While ATMs and the handiness of cashback at Sainsbury's are a relatively new phenomenon (as is holding on the telephone to speak to someone in another country about your overdraft limit), banking is not. Dating back to 3000 BC, 'banks' in the form of guarded temples in Babylon stored farmers' deposits of precious metals, farming tools, grain and other produce. Loans of sorts were secured against these valuables and the receipts worked as money that could be exchanged for other goods and services. The providers of said goods and services could cash them in at the local temple.

The Romans developed the idea of banking, charging interest on loans and paying interest on deposits, but, with the rise of Christianity, money-lending for interest was increasingly deemed immoral and the practice of banking became distinctly un-trendy. It ended up that Jews (who were themselves excluded from many 'respectable' professions) turned to banking as a means to an end, for their religion allowed Jews to charge interest to gentiles (non-Jews).

The word bank is derived from *banco*, Italian for bench. When a Jewish lender went bust, the story goes that he would break his bench to let everyone know about it. Once done, he was left only with a *banca rotta* (broken bench), and this is the origin of the term 'bankrupt'.

The Bank of England dates back to 1694 and was founded by Scotsman William Paterson. In exchange for the right to issue banknotes, the Bank lent the government of the time £1,200,000 – quite an overdraft in those days!

Barcode

It was a little bit of eavesdropping that led to the development of the barcode. At the end of the 1940s, an American postgraduate student named

Bernard Silver chanced to hear the president of a food-store chain requesting that the Dean of the Drexel Institute of Food Technology come up with a way of recording product information at shopping tills. He was turned away, but the conversation set Silver thinking, and soon enough he and his friend Norman Wood (another student) rose to the challenge by starting on some research.

Initially they thought fluorescent ink that would show up under ultraviolet light was the way to go, but it was too expensive and inefficient. Going back to the drawing board, they soon came up with a system inspired by Morse code. In place of dots and dashes, they used straight lines of different widths that could be scanned. They decided to arrange them in concentric circles for ease of scanning (the scanner could scan from any angle). The patent for a data code was filed for in 1949, with IBM eventually marketing the first usable commercial system in 1973 – the system remains today (note that the original idea of straight lines won out in the end), and those in the know refer to it as the Universal Product Code.

Cash Register

American bar owner James Ritty had a problem. His bar was a popular one but, despite all the beer being

183

sold, much of the money was not finding its way into his pockets. Rather, a significant amount of cash was ending up in the hands of his pilfering bar staff. In 1878, on the verge of a mini-breakdown over this, James took himself on a cruise to Europe to relax and think. While in the steamship's engine room, he observed a machine that recorded the revolutions of the propeller shaft, and this set him up with the idea for a machine that would count and store cash. The idea became a reality once James and brother John got together back home to devise a machine with a dial and keyboard that showed a total figure for sales each day. What's more, it had a bell that rang to let John know whenever it was being used by his staff. They tinkered with the design, dispensing with the dial, and in 1879 patented a version with tabs that popped up to show different amounts. Ritty's idea was bought by coal merchant John Patterson, who formed the National Cash Register Company in 1884 and set about making a fortune.

Credit Card

The concept of the credit card was first introduced by American hotel chains, who gave their customers 'shopper's plates' – payment cards that could be used in any of their branches. But it should be noted the cards could only be used to purchase services from one company. The man who came up with the idea for a card to be used in a range of businesses was Frank McNamara, an American businessman. The story goes that his idea was born when, having entertained a group of friends at a restaurant, the bill arrived and Frank realised he'd left his wallet elsewhere. As he was a regular customer, the restaurant let him leave his business card as ID so that he could come back and pay later, and this gave Frank the idea to create a card that could be used in many restaurants to allow diners to pay later on. Out of this the Diners Club card was born in 1950. It could be used in 27 New York establishments and worked very simply: Diners Club paid the bill, and the cardholder paid Diners Club.

The first credit card produced by a bank came along in 1958 – the Bank America card could be used for any type of purchase and did not have to be paid off every month (though interest was of course being charged on unpaid balances). This card was renamed VISA (Visa International Service Association) in 1976. The first British bank to launch its own card was Barclays who produced the Barclaycard in 1966.

Gift Voucher

The first coupons or vouchers came from an idea Benjamin Babbit dreamed up in 1865. Babbit was a soap-maker, a vocation that was considered quite an art back in those days. Before Babbit came along, soap was sold in slices from blocks in grocery stores, rather in the manner that cheese is still served in delicatessens today. Babbit came up with the idea of selling the first individually wrapped soaps as special luxury items. But he quickly realised that consumers would be deterred by the idea that they might be paying extra for the wrappers. To ensure against this, Babbit stamped the wrappers with a coupon and publicised the fact that ten coupons could be exchanged for gifts such as pictures. The idea took off and boosted sales of his wrapped soaps to the thousands in a very short space of time. So successful was Babbit's scheme that coupons were soon offered to customers for their loyalty on purchasing a large variety of products from the same shop. The idea for gift vouchers spread and grew from here, when they eventually became products sold in their own right.

Money

Banks that exchanged receipts in return for goods (see page 181) were around long before the existence of cash. Cash evolved out of this, together with the principles of exchange and barter. As it was often a little tricky to find things of the same value to exchange, tokens such as amber, beads and feathers became a primitive form of money around 700 BC. The use of coins came from the Middle East, specifically Lydia (eastern Turkey). Issued by King Gyges in 640 BC, they were roughly wrought pieces of metal with an impression of the royal crest stamped on to them. The Greeks were instrumental in developing the quality of coins, but it was the Romans who were crucial to the development of the function of money and a single currency. Interestingly, a major role of coins in Roman times was to communicate messages from the emperor to the general population. The messages were conveyed using symbolic and allegorical images on the back of the coins. Paper money began in China in AD 650 as coins were in short supply, but the first banknotes as we think of them did not come about until 1661 and were produced by the Swedish Riksbank. The Bank of England followed suit in 1695, although you wouldn't have seen folk queuing at cash machines until 1967, when Barclays unveiled the world's first cash dispenser in London.

Vending Machine

It was Greek inventor Hero of Alexandria who came up with the first known vending machine design in around AD 60. He conceived a device that, when a coin was popped into it, would automatically release a measure of holy water to the customer. Alas, the idea was never realised in his lifetime. The first 'coin-in-the-slot' vending machines came about in the form of honesty boxes, which became part of the furniture in English pubs in the early 1600s. A penny was inserted, releasing a lock on a box containing tobacco. Customers were trusted to take no more than a pipeful and then shut the box again, whereupon it would lock itself. These boxes were in use for over 200 years.

But it was only in the late 19th century that the first commercially successful automatic vending machine arrived on the scene. Located on the London Underground platform of Mansion House station in 1883, it was invented by Percival Everitt and dispensed postcards. The machine worked well, but, according to its creator, once in the public domain it fell victim to vandalism in the form of 'paper, orange peel and other rubbish' being 'maliciously placed into the slit provided for the admission of the coin'. So vandalism's nothing new, then! However, Percival persisted in promoting his product and eventually succeeded in making his vending machines a common sight around the country, with cigarettes, eggs, biscuits, towels, even accident insurance and much more besides available for purchase. The idea spread to America, and in the late 19th century the citizens of Utah could even buy divorce papers with a bit of loose change! Today, vending machines are more popular than ever, particularly in Japan where anything from fast food to used underpants (yes, used!) to live lobsters can be purchased at the drop of a coin.

CHAPTER ELEVEN:
INFORMATION, COMMUNICATION AND ENTERTAINMENT

'For goodness sake Logie, stop staring into space
and invent the television.'

Television

INFORMATION, COMMUNICATION AND ENTERTAINMENT

Calendar

As any calendar inventor will tell you, your basic problem when trying to invent a calendar is the fact that a solar year (the time the earth takes to go precisely once around the sun) does not contain a whole number of days or lunar months (the time it takes for the moon to change from new to full and back). Ancient calendars had a tendency to run fast or slow, because the 'year' denoted by them was always shorter than the solar year, and the lunar month averages 29.5 days.

All this created a real headache for the Egyptians, for example. Unlike other calendar-makers, who got themselves into all manner of trouble adding in months here and taking away days there to keep in step with the years, the Egyptians ignored the moon, instead using 12 months of precisely 30 days each, and threw in an extra five days at the end of the year which weren't part of any month. This worked up to a point, but, because their calendar gave a year that was a quarter day shorter than a solar year, they ended up with an extra 25 days every 100 years.

The Western, or Gregorian, Calendar uses a calendar month with a complete amount of days, 30 or 31, with 28 for February. For adjustments (because there are just under six hours a year left over), a 29th day is added to February every fourth year. When this happens it's known as a 'leap year'. This calendar is based on the Julian Calendar, instituted by Julius Caesar in 46 BC. Pope Gregory XIII made some adjustments in 1582. He eliminated the error that had accumulated over the years, and then restricted century leap years (years with an extra day) to those divisible by 400. It took other states a while to accept this new style, and Britain did not adopt the Gregorian Calendar until 1752, when the error amounted to 11 days. So, 3 September 1752 became 14 September, and at the same time the beginning of the year was put back from 25 March to 1 January. Russia did not accept the Gregorian Calendar until the October Revolution of 1917, so the event (which occurred on 25 October) is now celebrated on 7 November. What a brain-twister the history of the calendar is!

Calculator

For Egyptians with sums to do, there was nothing handier than the abacus, invented around 3000 BC. A real hit with the Chinese, it was used for many centuries. This was all very well, but it was William Shickard, a German, who invented the first mechanical counting machine in 1624. His 'calculator clock' was able to perform addition, subtraction, multiplication and division, but the first commercially viable calculator was the work of Thomas de Colmar in 1820. Over time, the design was much improved and his 'arithometer' was getting a lot of attention by the 1850s, and by the 1880s it was hugely popular, particularly with insurers. These machines were a long way from the pocket versions we know today, for which we have the American John Kilby to thank. He invented the microchip, and was very keen to put it to good use. Kilby and his pals at Texas Instruments – Jerry Merryman and James van Tassel – came up with a handheld machine in 1967. In 1970 Texas Instruments and Canon Inc launched the Pocketronic, the world's first electronic pocket calculator. To continue the history of the calculator's development would require the addition of several pages to this book, and there just isn't space. If you simply *have* to know what happened, go figure …

Camera

Ever wondered when the first 'Kodak moment' was? Well, it was in 1826, and was the result of the work of French inventor Nicephore Niepce. He spread a pewter sheet with some tar, inserted it into a box with a lens and positioned it in the window of his workroom. After eight hours, a permanent image had formed. But it was William Daguerre who developed the Daguerreotype in 1839, a commercial system that used silver iodide acting on a silvered copper plate. However, extremely long exposure times were required, a problem that was addressed by William Fox-Talbot in 1841. His calotype process used silver salt-soaked paper. Crucially, Fox Talbot realised that, before the salts visibly darkened the paper, a hidden image was formed. Moreover, this image could be revealed by the use of a developing solution, hence shorter exposure times. These paper negatives could then be used to make prints. Still, photography remained an expensive and complicated business, and anyone taking 'snaps' would have incurred great cost in developing them themselves.

In 1888, the Kodak camera, invented by George Eastman, changed all this. His camera came pre-loaded with film, and the user could return it to him to have the pictures developed and the film reloaded. Photos weren't the only things to be developed: so too was an excellent relationship between Eastman and his bank manager, as all this made George plenty of money.

The first 35mm camera was the Leica, the brainchild of by Oskar Barnack. First produced in 1924, it had been many years in the making and revolutionised the world of photography, as did the Polaroid camera, the idea for which came about in 1943. The story goes that it was the young daughter of Edwin Land, the inventor of the Polaroid, who catalysed its invention. Innocently asking why she could not view the picture her father had taken of her immediately it had been taken, three-year-old Jennifer set in motion a thought process in her father's mind that made him determined to find an answer. Land stated that within an hour he had the fundamental idea of how it would work, but the process of perfecting the idea took a further three years. The secret was in an exposed negative that passed through chemical-covered rollers and reacted with them to produce a brown and white print. The camera was launched in 1948 and achieved astonishing commercial success.

CD

As is the case for so many inventions, it was one man's frustration with the limits of current technology that fed the need to improve upon it. In the case of the CD, it was American James T Russell whose annoyance with the damage done to his precious vinyl records by the stylus made him dream of a disc capable of being read by a laser. He patented the idea in 1965, developing ideas for a system that transferred data from analogue to digital. But his funding ran out, and it was Philips who continued developing Russell's idea until they arrived at the video long player in 1969, a 30cm laser disc designed for visual images. The brainchild of Klaas Compaan and Piet Kramer of the Netherlands, its first public outing was at a demonstration in 1972, but it never caught on after its 1982 release. But all the while another Philips engineer had been working on a disc designed purely for audio, and it was this that led to a breakthrough. In 1982, Philips and Sony teamed up to launch the 120mm CD. The lead conductor of the Berlin Philharmonic Orchestra, Herbert von Karajan, helped launch the product, writing off all other forms of recorded music as 'just gaslight', and rumour has it that his influence over Philips was so strong that, at prototype stage, he had successfully encouraged them to change the CD from 115mm to 120mm so that it had enough for Beethoven's *Ninth Symphony*. One thing's for sure: the CD utterly revolutionised the way we listened to music.

Computer

The need for computers arose in the 18th century, largely due to two factors. Firstly, the laws of motion and gravity set out by Sir Isaac Newton resulted in scientists working with his theories having to perform hugely elaborate calculations. Secondly, those involved in new forms of commerce during the Industrial Revolution needed to make accurate calculations relating to a vast array of financial matters. For such industrialists, trusting the momentously huge logarithm tables (which had been compiled by teams of casually employed youths under the supervision of mathematicians) was simply no longer an option – there was no room for human error when money was involved!

In response to this, Charles Babbage came up with the idea for a mechanical device to perform such calculations, and the 1820 design for what he called a Difference Engine makes him the most oft-cited inventor of the computer. He then came up with the even more advanced Analytical Engine. However, due to their complexity (each comprising thousands of gears, cogs and axles), neither got past design stage and Babbage died in 1871 before anything resembling his concept was made.

Not before the 1930s did machines approaching the calculative abilities of Babbage's complex designs begin to appear. A vast 'analogue computer' called the Differential Analyser, designed by Scottish physicist Lord Kelvin, took up a vast room and was able to tackle a range of mathematical problems. However, it was not until binary code, which expresses numbers as a series of ones and zeros and is compatible with electrical machines, began to be used for electric computers that things started to move on in leaps and bounds. It is at this point that the development of computers becomes too complicated and specialist a story for a book of such general subject matter to fully recount! Suffice to say that, after various stages of advance, the first machine that fulfilled the criteria of a modern computer – programmable and with a memory able to store information – was designed in England at Manchester University in 1948. Known as a 'stored computer', its capabilities were limited, and it was another machine completed at Cambridge University the following year (the Electronic Delay Storage Automatic Computer (EDSAC) that was first adopted to aid research in the sciences.

From the 1950s, things moved onwards and upwards in various stages, but it was the 1969 programmable microchip (which squeezes thousands of transistors on to a tiny electronic circuit) produced by Intel that directly led to the microprocessor. It is this incredible piece of technology that eventually led to production of the home computers we have sitting on our desks (and laps) today, the first of which was the Apple II, launched in 1977 by Apple founder Steve Jobs.

Clock

The first mechanical clock was rather large – standing at 30ft, it was designed by Su Sung, from China, in 1088. It relied on a water wheel that periodically stopped to empty a bucket once it had filled, thus marking time intervals. Clockwork closer to our understanding of it was invented by Henry De Vick in the 1360s and hung in the Palais de Justice in Paris, but with only one hand and an error margin of up to two hours a day it was not as accurate as Su Sung's model. A damn sight smaller though! Timekeeping did not advance much in terms of accuracy until Dutch scientist Christian Huygens invented the pendulum clock in 1656, based on Galileo's principle of oscillating motion.

Despite these developments, for many years people relied largely on sundials and sand clocks, or hour glasses. Up to the late 16th century, these were the only clocks available for use at sea, and the Royal Navy used them right into the early 1800s.

Fax Machine

Believe it or not, the fax machine was invented over 150 years ago, and the principle laid out in 1843 by its creator, Alexander Bain, is still at the heart of the modern fax. A Scottish clockmaker, Bain took a couple of pendulums, linked them with a wire and got them swinging in sync with each other. The message to be transmitted was written in an electrically conductive material, and a piece of paper then wrapped around a rotating drum. A needle was attached to one of the pendulums, and as this pendulum swung over the drum the needle picked up electrical impulses and transferred them along the wire to the other pendulum, which had a pen attached to it. This pendulum then reproduced the message. Ingenious! However, Bain never managed to make a transmission, and the first fax was not sent until 1851, when English physicist Frederick Bakewell used an improved version of Bain's model in a demonstration at Britain's Great Exhibition.

The first commercial fax line was set up by the French government in 1865. It used a machine invented by Italian abbot Giovanni Caselli, who based his 'pantelegraph' on Bain and Bakewell's ideas. Napoleon III was a big fan, and supplied the necessary telegraph wires for the project to be implemented. The line ran from Paris to Lyon, but the job of popularising fax-machine use was tough thanks to stiff competition from the burgeoning telegraph industry. Big money was tied up in promoting the use of Morse code to pass messages, and people were inclined to believe the fax was only any good at transmitting images, not words. During his lifetime, Caselli saw the fax become virtually redundant before he died in 1891.

Merely a decade later in 1902, German scientist Dr Arthur Korn pioneered photoelectric scanning. Bain's original principle remained, the difference was that a photoelectric scanner mapped the black and white parts of the document, eradicating the need for electrical conductivity. By the end of the decade, the fax machine had become integral to newspaper offices around the world, but interestingly it was over 40 years before other areas of commerce caught up. In fact, it was not until the early 1960s that faxing became the norm in most offices. In modern faxes, images are digitised and broken into a grid of dots, the data is transmitted down the line and once at the other end translated so the dots can be reproduced as an image.

So there it is, all the facts about the fax.

Internet

If you type 'origin of internet' into Google, the results will give you some idea of how complex the story of its evolution is! However, it is possible to offer a potted history of it. The technology that allowed this modern phenomenon to become a reality goes back to the 1960s, when one computer could only be connected to another via a cable or a telephone line. Computer networks were made possible with the advent of 'packet switching', a process that converts data into units, or 'packets'. These can be passed on electronically to a central location. In turn, a 'connected' user can collect the information from that location through a 'router'.

The first significant network was the ARPANET (Advanced Research Projects Agency Network), a 1969 US Department of Defense project designed to remotely link up research computers. However, ARPANET ended up being primarily used by people to communicate between computers. So, effectively, this is when email was born. More and more networks emerged as different networks linked with one another over a period of years, thus creating a 'network of networks'. By the early 1980s this was known as the Internet.

While this allowed a huge amount of information on the net to become available, there was no easy way to find what you wanted unless you knew exactly where it was located. The solution to this problem was found by Englishman Tim Berners-Lee, the man credited with the creation of the World Wide Web. Working for CERN, the European Particle Physics Laboratory, he developed HTML (Hyper Text Mark-up Language), a universal code that instructs a computer how to present a web page, HTTP (Hyper

Text Transfer Protocol), which lets users and servers communicate, and URLs (Universal Resource Locators), which locate documents, images and files in a virtual world of information. This standardisation of the basic nuts and bolts of the net, plus the fact that Berners-Lee purposely left HTML, HTTP and URLs unpatented (keeping the technology freely available), is what led to the infinite resource that is such an intrinsic part of our lives today.

iPod

The iPod has become a must-have accessory for any music-loving urbanite, and outsells every other portable player on the market. In 2001, Apple launched it as the machine to put '1,000 songs in your pocket' and it has flown off the shelves ever since. The first models relied on the MP3 music compression system (invented in 1998), which removes unnecessary information from a file to reduce its size, thereby allowing many tracks to be stored on a relatively small memory space. Later on, Apple used its own more efficient format. Arguably, though, the success of the iPod is largely down to its sleek design – the only things contained in those pretty white cases are a hard drive and a computer to control it.

The name iPod was inspired by the film *2001: A Space Odyssey*, specifically the line 'Open the pod bay door, Hal!', spoken in reference to the USSC *Discovery One*'s white EVA Pods, designed for 'Extra-Vehicular Activity', or maintenance.

Mobile Phone

Only when we lose our mobiles do we remember what life was like before they were universally used – pretty similar! – and when nothing much changes in the absence of our little phones we may well start to wonder how necessary all the constant texting and phoning we do these days really is. But of course it's only a matter of hours before we've sorted out next-day delivery of a replacement – having probably agreed to a few more free minutes at 'knockdown prices' along the way ...

Although we think of them as a relatively recent invention, mobile phones have been around since the late 1940s, though the technology to make them work did not come about until the 1980s. Mobile radio, a mobile phone of sorts, was first used by the police in Prohibition-era America (1920–33) to aid their surveillance of gangsters. Dubbed The Untouchables, the force was headed up by Eliot Ness and used radio channels that eventually became controlled by the Federal Communications Commission. The commission dealt with the problem of overcrowded analogue channels by limiting their use to police and emergency services.

AT&T and Southwestern Bell were responsible for the first stages of commercial mobile communications, launching the first primitive service to Saint Louis, Missouri, in 1946. A push-button system allowed one person to speak at a time, and was linked to a switchboard via five receivers strategically placed around the city.

But the mobile as we think of it was the 1973 brainchild of Motorola's Dr Martin Cooper. Cooper famously used his Motorola DynaTac, the first mobile phone (about the size and weight of a brick!), to put in a call to arch-rival Joel Engel. At the time Engel was head of research at Bell Laboratories, who were also in the race to develop mobile technology, and Cooper used the phone call to let Engel know how swimmingly everything was going at Motorola! Given the nature of the call, we can assume it would not have lasted long, but had they wanted to the men wouldn't have been able to talk for more than half an hour, as that was roughly the talk time available on Cooper's early model, which took ten hours to charge!

Paper

The earliest proper paper was papyrus (from which the word paper originates) made out of water reeds taken from the Nile by the Ancient Egyptians. However, it was a luxury only available to the wealthy. Paper gradually became more accessible to the public from about AD 100 in China when developments were made by Cai Lun, an eminent eunuch in the court of Emperor He. It was Cai Lun who first began making pulp from cloth, ragging and fishing nets. Later he made pulp by mixing hemp, flax and water with tree bark (for bulk). The mulch was filtered by being forced through bamboo that had been split to make small holes. The filtrate was then left to dry and harden on a sheet. By the 10th century, paper-making methods leaked out of China and mass production began. Literacy was on the rise and paper demand was increasing correspondingly. In 1718, Louis Robert saved the day by inventing a machine that could knock out reams of paper on a constantly moving production line: meeting demand was no longer a headache. Since then, paper production increased exponentially to the point where in the late 1980s Britain's total paper usage was somewhere in the region of a whopping 10 million tonnes – a far cry from the few sheets available during the earlier part of the millennium.

Pen and Ink

Ink was originally concocted from smoke of pine, lamp oil and water-soluble substances from animal tendons approximately 5,000 years ago. Alternative recipes included dark berries, plant matter and metallic compounds. The ink we use today is made from a chain of chemical reactions and produced by industrial manufacturers.

From 300 BC, writing implements were made using large feather quills with slits in the tip for the ink. Quills were used right up until the mid-1900s when fountain pens came into popular use. Though the first fountain pen was developed in Paris in 1902 by Monsieur Bion, it wasn't patented until the 1880s when American Lewis Waterman designed an air-flow mechanism that altered the pressure and allowed a reliable flow. Only then did they come into common use. Nearly 40 years later the possibility of mass-producing pens appeared on the horizon when Laszlo Biro developed the ballpoint pen out of pure frustration. Being a journalist, he became tired of the ink drying up in his pen or, conversely, wetting his page. He observed that the ink used for printing the journals dried more quickly so he thought about ways to release the superior ink from a pen in a more controlled way. His favourite design was a metal ball, placed in the fine tip of a pen, that could roll across the page neatly. He first patented the ballpoint pen in 1938 in Argentina, and the Biro was sold under its inventor's name. Marcel Biche designed his own ballpoint that was suitable for mass-production and which is still selling more than any other make – over 10 million pens per day worldwide!

Pencil

The first pencil-like writing implements used were simply pieces of hard reed used for inscribing clay. According to the *Webster's Etymological Dictionary*, the word 'pencil' comes from the latin *penecillus* meaning paintbrush or 'little tail', and came into use in the mid-16th century. Around this time a large, abundant supply of the mineral graphite was found in Borrowdale, Cumbria, and this is where the first true pencils were made (visit the pencil museum at Borrowdale to find out more!). Graphite is a naturally occurring soft, shiny element that marks things easily, which made it ideal for fashioning writing implements from. Crude pencils were made by slicing the graphite into thin rods that were held in a protective case or holder. Later on, string was wrapped around the rods to create a holder, and in 1812 graphite sticks were held firmly in place inside a wooden rod. As graphite looked like black lead, it was mistakenly called plumbago and this is the reason that people still talk about lead pencils. Due to the abundance of graphite in Cumbria, England held the biggest chunk of the market. In fact, such a large single source of graphite has never been found since. Other countries produced pencils using less pure local graphite, which resulted in a poorer product. It wasn't until one of Napoleon's officers hit on combining graphite with clay and then baking it in a furnace that quality pencil-making without pure graphite hit mass-production levels.

Postage

Well before Postman Pat began his rounds or Wish You Were Here postcards were being sent by holidaymakers, the Ancient Egyptians were developing highly efficient postal networks. Around 2000 BC, messengers were spaced regularly along well-trodden routes and kept very busy relaying important messages back and forth. This practice was adopted for a while by the Roman Empire, but died with it. Although private postal services existed to serve medieval dignitaries and clergymen, government-run postal services only emerged from the 15th century onwards. A service was established in France in the mid-1400s, and Henry VIII famously set up the Royal Mail in 1512, but only for government post. Public post arrived in Britain in 1635, and by 1680 the London Penny Post, conceived by William Dockwra, used a pre-paid hand stamp. The stamp went some way to aiding the rather cumbersome system of charging for postage according to distance travelled. The flat charge worked a treat locally, but did not work between towns and cities.

The stroke of genius that solved this problem came from Rowland Hill in 1837. In his 'Post-Office Reform' paper, he observed that it was costing a fortune to calculate postage fees based on distance, more indeed than the cost of transporting the letters! Hill advised the government they would be much better off with a standard charge, whatever the distance. They wisely listened and in 1840 the Penny Black, Britain's first postage stamp, was issued.

Radio
Radio waves occur naturally, so were not 'invented' as such, but it was in 1887 that German Heinrich

Hertz proved the existence of what Scotsman James Maxwell had theorised earlier – that a remote effect could be caused by electricity. Despite his discovery, Hertz did not see any purpose for this effect. It took the experiments of Italian Guglielmo Marconi to successfully harness 'electromagnetic propagation', more commonly known as radio waves, for commercial communication. Working on improving the Branly tube, a device used to detect radio waves, Marconi invented the radio antenna, and experimented further with the assistance of his brother Alfonso. They transmitted Morse code signals to each other across a field – Alfonso would acknowledge receipt of a transmission by waving a flag. When Marconi moved on to check if signals could pass over hills and mountains, Alfonso was out of sight and fired a hunting rifle into the air to let Guglielmo know he'd got the message.

Despite the monumental implications of his experiments, the Italian government showed no interest in helping Marconi to develop them, so he decamped to Britain. Unfortunately, British customs officials went through his bags and managed to break his radio equipment, labelling it an 'infernal machine'! All was not lost, however, and Marconi went on to give the first public demonstration of 'wireless' in 1896 in Toynbee Hall, London. The next year he set up the Wireless Telegraph and Signal Company, which manufactured radio equipment and maintained radio stations. In 1909, he received a Nobel Prize for Physics for his efforts in the field of wireless technology. Interestingly though, the 'father of radio' turned out not to be the true father, and in 1943 Serbian inventor Nikola Tesla was credited as radio's true inventor. Marconi's patent was overturned in favour of Tesla in the US Supreme Court – his Tesla Coil had been patented before Marconi's experiments, but Marconi's influential friends had managed to get the US patent for radio taken from Tesla and given to Marconi in 1904. Alas, Tesla died in 1943, just before he would have seen the patent rightfully returned to him.

The first radio broadcast took place on Christmas Eve 1906. Canadian Reginald Aubrey Fessenden used a 420ft-high radio mast in Massachusetts to broadcast himself playing Gounod's 'O, Holy Night' on the violin. He also sang some songs and read a few passages from the Bible. His transmission was picked up by ships at sea. A year later, the first British broadcast occurred when Quentin Crauford of the Royal Navy aired a concert from HMS *Andromeda* for the benefit of other ships. But it was once again Marconi who was responsible for the first broadcasting service in Britain in 1920, producing 30-minute programmes that went on air twice a day. Eventually, along with a couple of other stations, Marconi's 1922 station 2LO was bought up to form the British Broadcasting Corporation (BBC) in the same year, when the first BBC news bulletin went out on 14 November.

A further significant development in radio occurred in 1934 when Edwin Armstrong introduced FM radio. 'Frequency modulation' uses more complex technology than 'amplitude modulation' (AM) – the frequency transmitted adjusts itself to suit the sound wave it is carrying, helping to avoid interference from other electrical equipment.

Telephone

Alexander Graham Bell was a man obsessed with communication. Fathered by a speech teacher who had devised a system to help deaf people speak, Alexander was immersed in conversations about speaking and hearing from an early age. When he grew up, Bell junior moved from Scotland to Canada and established a school to train teachers of the deaf. He began experimenting with harmonic telegraphs, devices that used musical notes to relay messages. He observed that, when close enough to an electromagnet, a strip of iron would replicate vibrations from another strip that was connected by wire. This set bells ringing in Bell's head. He wondered if he might be able to use the principle for the transmission of sound, particularly speech.

After many experiments by Bell, helped by his pal Thomas Watson (a mechanic), a device to perform the task of transmitting speech was arrived at. Essentially, it worked as follows. Two diaphragms were connected by wire. When spoken into, one diaphragm, which was attached to a wire dipped into acid held in a metal vessel, caused the wire to move in the liquid, which changed its resistance. This produced a 'varying current' that was conveyed down the line by a receiver, causing the other diaphragm to vibrate, replicating the original sound in a reverse process. Simple? Bell wasn't so sure, and always asserted that he didn't really understand how his machine worked, but was just pleased that it did!

The first 'telephone call' was unintentional and only seven words long – 'Mr Watson, come here, I want you' were the words uttered by Bell from his room as he tinkered with his transmitter. By coincidence, Mr Watson was having a play with his receiver in another room, and heard the words emitted from it. This was the moment they realised they had cracked it. The date was 10 March 1876.

Worth knowing is that the story of the telephone (from the Greek *tele*, meaning far off, and *phone*, meaning sound) could have been very different. Alexander Bell filed for the patent for his telephone only hours before rival inventor Elisha Gray was planning on turning up at the Patent Office with a virtually identical idea! As soon as Bell started producing telephones to huge commercial gain, he faced nearly 600 separate law suits from Gray, who claimed that Bell's phone used ideas developed by Gray. Gray attempted to prove this by showing that ideas were being used that were absent from Bell's original patent. However, all suits brought against Bell were unsuccessful.

Television

It was John Logie Baird in 1924 who worked out how to put together the necessary components for a working television. He must have known he was on to something good because he filed his patent two years before presenting it to the public in 1926. Baird didn't have wealth on his side, but he still managed to make a working model. This comprised a mechanically functioning transmitter and receiver combined with an optical scanning disc (previously invented by Pail Nipkow in 1884) in combination with scrap metal taken from biscuit tins, knitting needles and piano wire: it produced images using telegraphy in 1925. To test his invention he grabbed a young boy from the street and plonked him in front of the transmitter. To Baird's great disappointment there was no image on the screen in the other room. He then realised that the boy was scared of the white light emitted from the transmitter and had jumped to the side of it. He pleaded with the boy to stay glued to the spot and even slipped him half a crown for his trouble. He fled to the other room and, lo and behold, an image of the boy was clearly projected on to the screen. He presented the first working television in London the following year.

Baird may have been the first to have his idea accepted by the public, but he wasn't necessarily the most important man in the history of the television. Many would argue that this accolade goes to Vladimir Zworykin, whose work produced the electronic television that superseded Baird's mechanical version. Zworykin's was a cathode ray model, and used a cathode transmitter and receiver. It was these televisions that became essential kit for couch potatoes all over the world.

Thermometer

The word thermometer comes from Greek, *therm* meaning heat and *meter* meaning to measure. In 1714 Gabriel Fahrenheit invented the first mercury thermometer. In response to a temperature increase, mercury was found to expand in volume and move up the column proportionally and show a temperature change on a scale. Fahrenheit drew fixed points on the scale and marked the temperature at which water freezes as 32°F, and that at which water boils at 212°F. Approximately 40 years later Anders Celsius sought to improve and standardise the mercury thermometer. His version possessed a scale with the boiling point of water at zero and the freezing point of water at 100 degrees; later he reversed these points and put freezing point at zero and the boiling point at 100, the scale we know today. The gradations were termed degrees Celsius (°C) and the scale the 'centigrade scale', and this became the official scale for scientific purposes, replacing the Fahrenheit scale which is now only used in weather terms.

Video Recorder

From the 1920s onwards several attempts were made to develop video recording machines. Innovations using wax discs were tried but these were problematic due to low durability and the propensity to cause distortion. This limited success continued until Charles Ginsburg and Ray Dolby thought long and hard about the problem of video signals, focusing in particular on the fact that they are of such high frequency. Using an ordinary tape recorder didn't suffice as it was not fast enough to pick up television signals, and speeding the tape up did not solve the problem either. They pondered further and tried running the tape at the usual speed while scanning the tape at a much higher speed to record the television signals. The result was a longer piece of film packed into a shorter tape. At first only television companies could afford such novel equipment, but before long electrical companies raced to perfect the new machine for domestic use. In 1972 Philips began selling the first machine, which became known as the 'VCR', or Video Cassette Recorder.

Watch

The first watch was slightly cumbersome – about the size of a circular beer mat. In 1500, it was produced by German locksmith Peter Henlein, who, according to Johannes Cocclaeus, a Nuremberg dweller at the time '[made] things which astonish[ed] the most learned mathematicians', including a watch that could 'strike for 40 hours', named the 'Nuremburg Egg'. Henlein's creation relied on a spring, rather than weights, to drive the clockwork. Watch technology developed from Henlein's invention, but the first clocks were only accurate to within about 30 minutes every day. As design improved over the centuries, so did accuracy, with minute and second hands eventually being introduced. It was not until the late 1800s that Abraham Breguet came up with the layout we know today, and, by the end of the century, the price of watches had begun to fall to the point where most people could afford them and it became a common item. Watches were not worn on the wrist until the early 1900s, and even then only by women. Only after the First World War, during which watches were worn by soldiers, did the notion that watches were effeminate slip away and the wristwatch become accepted by all.

Some decades later, the development of digital wristwatches was perceived by the watch-making industry as a threat, and lead to the use of quartz for improved accuracy in traditional styles of watch. Even though the cheap digital watches of the 1970s onwards were hugely popular, the analogue watch lives on today, and the popularity of the digital watch is fading fast. After all, every mobile phone has one …

Whistle

The whistle came about long before football referees had a need for them. With examples dating back to around 10,000 BC, it is thought that the first stage of whistle development would have been someone blowing over the end of some bamboo bone in the way we might blow across the top of a bottle. Things just went from there. In China, archaeologists have unearthed whistles containing more than one note that date back 9,000 years.

CHAPTER TWELVE:
INDOORS, OUTDOORS

'Trust us to inherit a bone idle poltergeist.'

Elevator

INDOORS, OUTDOORS

Air Conditioning

Air conditioning in one form or another has existed since ancient times. In Babylon, residents doused their houses in water, which drew heat from them through the evaporation process. Other ancient peoples draped wet leaves and cloths across doorways so that the air flowing into their rooms was cooler. So the basic idea of replacing hot air with cool was nothing new in the early 1800s, when fans began to be employed to cool air by blowing it across ice. This bright idea came from John Gorrie, a doctor who hung the ice from hospital ceilings under which lay patients struck with yellow fever. Although it worked, Gorrie's method did not take off. It wasn't until the early 1900s that modern air conditioning came about, thanks to one Willis Haviland Carrier. Taken on by the Buffalo Forge Company for his electrical engineering know-how, he was presented with the task of saving on the company's heating bill by working out the amount of heat absorbed by air passed over a heating coil system. He succeeded in cutting those bills, and moved on to work for a printing company who had a spot of bother with their paper going soggy on hot and sticky days – a printer's nightmare! They needed a constant temperature in the room, whatever the weather, and Carrier soon developed an ingenious machine that was up to the task – his 'apparatus for treating air' was a massive unit that heated or cooled a room. A fan drew the air in and either hot or cold water was sprayed on to and removed from it. The hot or cold air poured from the unit to regulate room temperature and humidity. His principle is now used across the world to keep people nice and cold in the office and save them the trouble of opening the windows …

Barbed Wire

This book is about the objects in our lives, and, while barbed wire certainly fits the bill in as much as it is often seen, it is hoped that the reader does not have the misfortune to come into contact with it too regularly. For, as we all know, it is nasty stuff. Nasty stuff that helped settlers get a hold on the American West to keep unwanted people out of settlements, and to keep cattle in (as the frontiers expanded there was a shortage of timber fencing for the job). Later, barbed wire became associated with imprisonment and much worse during the world wars.

But its origins lie in farming. The idea for barbed wire was patented in America in 1853, but the intricacy of design made the manufacturing process an expensive headache. It was not until 1873 that farmer Joseph Glidden worked out how to make it cheaply using a machine. His Barb Fence Company went on to make him a fortune. There are over 1,500 types of barbed wire, and some people are interested enough to actually collect the stuff, with rare pieces fetching hundreds of dollars …

Brick

Many of the houses in which we live, the offices in which we work and the bus stops that shelter us from the rain started with the single brick. The brick is the unit of robust building material that, when stacked, one on top of the other in the appropriate dimensions, can result in a sturdy structure able to withstand all manner of trauma.

Bricks have been around for at least 7,000 years. They have pretty much remained the same in appearance, but more sophisticated fabrication has led to a more durable material for construction.

Early bricks appeared circa 5000 BC when mud was strengthened with straw, shaped into blocks and dried in the sun. The problem here was that extreme weather conditions such as cold and rain could reverse the process and cause decay. About 1,500 years later in Ancient Babylonia, clay was collected from the Euphrates river and baked with roaring flames in kilns for a more hardy and waterproof product. The large amounts of fuel required for this more sophisticated brick made the process expensive, and therefore priority dictated how they were used. Though bricks were in high demand, they first had to be used for public pavements and walls. Only later were they used for permanent buildings such as temples and later individual homes. Earlier structures were not secured with anything between the bricks, so entire structures were improved in sturdiness when bricks started to be fixed. Very early fixing material was composed of straw, reeds and other waste material. Later, these materials were superseded by hard-setting cement or mortar made from mixtures of lime and sand.

Concrete

The remains of a hut dating from 5600 BC exist with a floor made of red lime, sand and gravel, and the Shaanxi pyramids in China, built thousands of years ago, contain a mixture of lime and volcanic ash or clay.

The 'recipe' for concrete hasn't changed much since it was first used in Egyptian times. Around 2500 BC lime and gypsum cement was used in the construction of the pyramids to keep stone blocks in their place. But it was the Romans who perfected the art of producing concrete as we think of it, using lime, volcanic ash and aggregate, often pumice stone. Today we use pebbles and powdered brick as aggregate, and recycled material is increasingly incorporated into the mix.

Reinforced concrete was invented by Parisian Joseph Monier in 1849 by pushing steel rods into concrete before it set.

Elevator

Louis XV of France had a mistress, who lived on the second floor of his private apartments. Now, whether it was out of laziness or discretion is uncertain, but King Louis did not want to climb the stairs to see her! Instead, in 1743 he had a 'flying chair' installed so His Majesty could be transported up to Madame de Châteauroux whenever he fancied. Although it was outside the building, the chair was within a private courtyard and served the king's balcony. A delicately organised set of weights were arranged in one of the chimneys to act as counterweights for easy raising and lowering by hand. This is considered to be the world's first 'passenger elevator'.

Elevators were commonly used in 19th-century industry, driven by either steam or hydraulics, and the only reason passenger elevators weren't in place was public uncertainty about their safety. To overcome this proletariat nervousness, American Elisha Otis provided a grand display when publicly demonstrating his 'safety lift' in New York in 1854. To the initial horror, and then amazement, of the crowd, he stood in his lift and requested the rope holding it up be chopped by an axe. As the rope dropped away, the lift stayed still, thanks to spring-loaded ratchets which quickly engaged the lift to the shaft in emergencies. The public never looked back, and nor did the lift industry.

Escalator

The first form of escalator was the Reno 'inclined elevator', designed by New Yorker Jesse Reno in 1892 and installed as a ride on the Old Iron Pier on Coney Island. Consisting of an inclined conveyor belt constructed out of wooden slats with forward-pointing rubber cleats attached for passenger feet to grip, it was driven by an electric motor. Despite a top speed of only 1½mph, it caused huge excitement in its riders, some 70,000 of whom gripped its handrails and grinned as they travelled up it. Harrods of London was the first shop to install one, at the behest of manager Richard Burbage. Able to transport 4,000 passengers an hour, it was a huge success. The trouble was that some customers were a little overcome by the novelty of the ride. To deal with this, attendants were on hand to dish out a brandy to anyone who needed it on reaching the top!

The first escalator with stairs was designed by American Charles Wheeler, an inventor who sold his patent to Charles Seeberger in 1898. Seeberger added his own improvements and teamed up with the Otis Elevator Company (see page 232). In 1900, they exhibited it at the Paris Exhibition and, just a year later, the first Seeberger escalator was erected at Gimbel's Department Store in New York.

Fire Brigade

Organised firefighting goes back to Roman times, when their *vigiles* (which loosely translates as 'firemen') used bronze water pumps to protect the public from the ravages of fire. This free public service died with the Roman Empire. It was not until much later that the idea of the fire brigade resurfaced. The first fire engines emerged in 1518, and fire hoses came along in 1672, making firefighting easier and safer. In the beginning, fire services were not public; rather, they were set up by fire-insurance companies. Originating in Germany, such organisations charged a fee to members, which was used to replace losses in the event of fire. Only owners of properties who were paid-up members benefited from the service. Realising that it would also be in their interests to fight the fires, thus reducing payouts, fire brigades were employed to do just this. Eventually, there were so many fire brigades that local authorities took over and merged them to form public brigades. The first large-scale brigade in England was the Metropolitan Fire Brigade, set up in London in 1866. Before that time uninsured buildings had been left to burn!

Fire Extinguisher

The first fire extinguisher came in the form of glass balls containing a saline solution. The idea was that the balls were thrown at a fire. Invented in 1734 by German M Fuches, the adverts for the balls were notable for featuring an entire family gathered together, flinging them around their burning drawing room with looks of rapture on their faces! These adverts appeared in Britain until the early 1900s.

Automatic fire extinguishers closer to the ones we know today were pioneered by barrack master Captain George Murphy in 1813 after he had witnessed a terrible blaze in an Edinburgh tenement block. The fire was so high in the building as to be out of reach of the hose, so nothing could be done to douse the flames. He felt that having small containers of water placed strategically around buildings could work wonders in stopping fires before they grew out of control, and his 1916 copper extinguisher held four gallons. Murphy's device contained water and pearl ash, as well as compressed air, and fire-extinguishing technology developed from here.

Glass

During the Stone Age, obsidian was the first type of naturally occurring glass to be used. It is a material that results from volcanic lava coming into contact with water. Gemstones were later made out of obsidian, and with a bit of elbow grease it could be polished up and used as a mirror. It was also a material out of which weapons could be fashioned.

The formula for making glass, so the story goes, was discovered pre-3000 BC. It is said that a merchant ship anchored for the night and the crew set about cooking their dinner on a roaring fire on the beach. When they couldn't find rocks to support the pots of food, they took lumps of nitrum (soda and potash) from the ship. On contact with heat and sand the nitrum formed a clear liquid that set into glass. Egyptians developed the method to a useful end in 1500 BC when they began pouring the molten liquid into moulds to make receptacles such as bottles and jars.

The Romans made the first glass panes for windows, though they didn't appear as they do today because the finished surface was very uneven, and light shone through it only very poorly. Still, the windows were of some use for those who could afford them, as they allowed a bit more light into their homes and some light was better than none at all.

In 1905, Eduoard Benedictus was the man behind the ingenious idea that led to laminated glass for car windows. Previously, standard glass would shatter in high-speed collisions and produce potentially lethal shards. Benedictus simply glued a very thin sheet of plastic between two thin sheets of glass. If the glass broke, the pieces would remain adhered to the plastic sheet.

Prior to 1952 the quality of glass was still poor. Window panes would be unevenly set and contained bubbles that caused unwanted light distortion. Flat surfaces could only be obtained by shaving off the uneven bits – a very laborious task. Alistair Pilkington tinkered with ideas to solve this problem for seven years, eventually cracking the glass question with a technique that involved heating the glass to a molten state and pouring a layer of it over liquefied tin. The tin served to flatten one side, and the other side was flamed to even it out. This became known as the float glass technique, and Pilkington Glass is still going strong today.

Hosepipe

The English country garden could not have reached such perfection without a readily available water supply delivered from the hosepipe to cover every flower bed and privet hedge with its fine water spray. More importantly, the trusty hose has helped put out many a potentially devastating fire. The first hosepipe appeared around 400 BC. Tubing was innovatively fashioned out of ox intestines, and large water bags were connected to it. The water flow was caused by somebody taking a running jump on to the bag. Later, in the late 1600s, piping was made out of leather stitched together in an attempt to strengthen hoses for use by Dutch firemen. Not so flexible and a whole lot more likely to burst, leather pipes were replaced by hoses made by a rubber company in Ohio, Goodrich. These superior pipes were made watertight with heavy-duty cotton stitching and Goodrich sold them to fire fighters the world over. As with many of the best inventions, the hosepipe had a secondary use which became clear when unmanageable orders for hosepipes started coming in from horticulturalists and family gardeners keen to water their flowers.

Lawnmower
In 1830 it struck Edwin Budding, a textile-factory worker, that rotary blades used for cutting large velvet cloths could be incorporated into a machine

that could cut grass. He teamed up with John Ferrebee to design the first lawnmower and they rushed to acquire the patent. However, they did not want to prevent other companies from manufacturing them. In fact, they were more than happy for this to happen – they were desperate for the patent because they simply wanted to maintain control over the idea, thus ensuring a percentage of all revenues from lawnmower companies who could apply for a licence to manufacture Budding and Ferrebee's machines. Their sales spiel was 'country gentlemen may find in using my machine themselves an amusing, useful and healthy exercise'. By about 1840 a machine was designed that could be pulled by a horse: the only snag was that the horses had to wear leather booties to save the well-kept lawns!

The 1850s saw the emergence of roller-manoeuvred mowers. Many lawns were mown using such devices until the end of the 19th century, when James Sumner revealed his design for the first steam-powered machine, fuelled by kerosene, and this was the birth of the modern mower.

Skyscraper

The first skyscraper was designed by William Jenney, an American engineer. Pioneering the use of steel girders, which are necessary for really tall buildings because they take the weight away from the brick or stone, his ten-storey building was built in 1885. Another important factor in the development of skyscrapers was the development of the safety elevator (see page 232), without which such tall buildings would have only been useful to workers with Herculean stair-climbing skills!

Smoke Alarm

Who would have thought that butter could be used to alert someone to the presence of smoke? George Andrew Darby, that's who. His 1902 device made ingenious (not to say eccentric) use of a slab of butter wedged between two metal plates – if temperatures rose to perilous levels, the butter would melt, causing the metal plates to come into contact, thus closing an electrical circuit and sounding an alarm. Strictly speaking, though, this was a heat detector, not a smoke detector. Smoke detectors rely upon light or low-level radioactivity in the detection of smoke particles resulting from the beginnings of a fire. The first alarm for home use went on the market in 1969. It relied on batteries, not butter.

Thermostat

Used in irons, kettles, ovens and central heating (to name but a few) to regulate and maintain a constant temperature by switching heating and cooling systems on and off, the thermostat was conceived and made by Scottish inventor Andrew Ure in 1830. Derived from the Greek *therme* (meaning 'heat') and *states* (meaning 'make stand'), it was patented as 'apparatus for regulating temperature'. The most important of his devices used a bimetallic strip which consists of two different metals that expand and contract under different temperatures. The movement of the strip causes a temperature controller to be switched on or off. Simple.

Wheelbarrow

Sometimes the genius of a great invention is its simplicity. The wheelbarrow is indeed a humble little invention, yet it has served with great usefulness on many occasions throughout history. Whether it be to shift heavy and awkward building materials, supplies for troops, or even to transport a load of shopping from the boot of the car, the wheelbarrow has proved indispensable for centuries.

Very early barrows were simply carts on wheels. These are reported to have been used in Greece in approximately 500 BC, but the consensus is that the wheelbarrow was invented in China around AD 200 by Chuko Liang to transport military supplies more efficiently. Liang tried to mechanically emulate the way in which oxen were used in the transportation of heavy goods. The fundamental principles behind modern-day wheelbarrows is not a million miles away from the original design. However, thanks to 13th-century European improvements, the ease with which they can now be manoeuvred is vastly superior. Whereas early wheelbarrows had two wheels, modern versions have one wheel and longer arms for leverage.

The advent of the wheelbarrow had a universally dramatic effect on labour costs, and freed up workforces to carry out tasks with hugely increased efficiency. The wheel was integral to every barrow until James Dyson managed to improve upon the seemingly unimprovable with his award-winning Ballbarrow in 1977. The wheel was replaced by a large plastic ball, making movement even easier and greatly reducing damage done to gardens by track marks from the thinner wheels of traditional models.

Just think, without the wheelbarrow, many of the cathedrals and buildings that were constructed before the age of machines might never have been completed!

CHAPTER THIRTEEN:
TRAVEL AND TRANSPORTATION

Parking Meter

TRAVEL AND
TRANSPORTATION

Atlas

Many cartographers tried their hand at mapping the world as they knew it, but the first publication that resembled what we would call an atlas did not come along until 1477. Published in Bologna, it contained 27 maps and was based on calculations made by geographer Claudius Ptolemy, who worked in Alexandria around AD 150. It has not been established if the maps were simply engravings reproduced from Ptolemy's originals, or if Greek scholars constructed them working simply from his text.

The first modern atlas emerged in 1570 to huge acclaim. Abraham Ortelius issued the *Theatrum Orbis*

Terrarum, a work of 53 sheets comprising every known country in the world. The word 'atlas' was first used to refer to a book of maps in 1595, when Gerardus Mercator printed his atlas. All early atlases had a frontispiece showing the Greek Titan Atlas supporting the globe, which explains why they were so named. Christopher Saxten produced the first English atlas showing the English and Welsh counties in 1579.

Aeroplane

The human desire to be up with the birds is ages old. Many will remember having their imagination captured by the Greek myth of Icarus, who managed to fly with a pair of wings strapped to his arms. The trouble was that the wings, which were made of feathers and wax, fell apart when Icarus got too close to the sun. Alas, poor little Icarus plunged to his death. For many, the idea of flying was always mere fantasy, but not for a certain second-century monk, Brother Elmer, who took inspiration from the plight of Icarus and set about making a primitive paraglider from wood and linen. He threw himself from a cliff and managed to glide 200 metres before breaking his legs in a crash landing, making him the first person ever to fly. Brother

Elmer was keen to give it another go, but the abbot at his monastery wasn't having any of it, so he spent the rest of his life on solid ground.

Centuries later, in the late 1700s, the aristocrat and lawyer George Cayley laid down the basic design of the modern aeroplane, with fixed wings and a tail attached to a fuselage. His glider was launched in 1853, but it was German inventor Otto Lilienhal who was the first man to build workable gliders that were heavier than air. After clocking up thousands of air miles, he came to a sticky end, fatally crashing one of his machines in 1896.

Powered flight was not far behind. In 1903, the Wright Brothers successfully launched *Kitty Hawk* with a flying extravaganza that lasted 12 seconds! Having nailed the issue of how to get a machine to take off and be controllable in the air, they truly have a claim to being the inventors of the aeroplane, and in 1905 they perfected their design with the world's first working model.

Airbag

You might think that the airbag is a relatively new 'part' to a car, and it is true that they only became standard in the last two decades. But the concept of air-filled devices was originally employed in aircraft in the 1940s – air-, nitrogen- or carbon monoxide-filled bladders constituted the first cumbersome devices. The modern airbag design was patented by John W Hetrick in 1953, a naval man who was concerned for his family's safety, and in 1968 American inventor Allen Breed finished the job by developing the ball-in-tube sensor, a device that enabled efficient detection of crashes. Initially, airbags were incorporated into cars as an alternative to seatbelts at a time when Americans hardly bothered using them, and General Motors were the first to introduce the 'Air Cushion Restraint System' in their 1973 Oldsmobile Tornado. Subsequent airbag-equipped models followed, but design teething problems caused a number of fatalities, so it was back to the drawing board. In the 1980s, Mercedes Benz introduced them to their cars, not as alternatives to seatbelts, but as accompanying safety devices. Since the mid-1990s they have become standard in most cars. These days they are very safe thanks to a venting system that makes the inflated bag less rigid. Thankfully, most of us never get to see them doing their job.

Bicycle

Freddie Mercury might well have sung 'I *don't* want to ride my bicycle' had he been around when the first model was invented in the early 19th century, as it didn't even have pedals! Named the 'hobby horse', and invented by a German, Baron Von Drais, in 1818, the wooden construct relied on the rider's feet pushing on the ground to propel it. It didn't really catch on, but in 1839 provided inspiration for Kirkpatrick Macmillan, whose invention may have pleased Freddie Mercury a little more. His model, once again made of wood (and iron tyres), included 'pedals' that connected to the rear wheels – a forward and backward movement of the rider's feet, rather like the action of wiping one's feet on a doormat, turned the wheel and set the machine along the road. Macmillan's sole motivation for inventing the bicycle was to get himself around, and he didn't give a thought to establishing his design commercially. Wherever he pedalled, crowds would gather in amazement. On one occasion, the throng was so strong that the cycling Scotsman accidentally knocked over a little girl, for which he was taken to court and fined five shillings.

The first bike with modern-style pedals was the 1867 'boneshaker'. Turning pedals were attached to the front wheel. Invented by Pierre Lallement, the 'velocopide' was far from a comfy ride, but it was nevertheless the first bike to undergo successful mass-production by Pierre Michaux in 1861.

A British invention followed: the Penny Farthing, whose massive back wheel set the rider perilously high up and led to many accidents, was so named because the smallest coin at the time was the penny, and the largest the farthing. Though they worked to a degree, they were certainly no good for 'wheelies', 'bunny hops' or 'off-roading' – there was still work to be done.

The first invention that resembled today's bikes was designed and made in 1885 by John Starley, an engineer. Following on from precursory 'safety bicycles' with chains, he was the first to make both wheels the same size. His bike was called the Rover Safety, and its design formed the basis of modern frame design.

Boats and Ships

Unsurprisingly, the first boats were simple canoes, or 'dugouts' – wooden logs shaped into floating vessels and used with paddles. The earliest known example is the Pese canoe, a 10,000-year-old craft discovered in Holland. Tools invented during the

Bronze and Iron ages allowed for quite advanced models to be constructed – several dugouts could be bound together to form larger vessels. Around 4000 BC, the first 'non-dugout' boats, built using planks and bundles of reeds, were made in Ancient Egypt. Archaeologists found 14 large vessels near Cairo. Built 5,000 years ago – planks were bound together using rope, and the reeds were used to make them as watertight as possible – they were more than likely buried with a pharaoh for him to use in the afterlife, this being one of the ways in which boats were used symbolically at the time.

It was around 2000 BC that boats we might call ships emerged. An example of a large vessel that would have combined sail and paddle power to carry many men has been found in Egypt. Its ability to cross open water with ease classes it as ship-like. For many centuries, boats with sails ruled the waves.

After many years of trial and error, in 1783 the world's first steam-powered boat finally made its maiden voyage on the River Saône at Lyons, France. It was built by the Marquis de Jouffroy d'Abbans and was a huge event in shipping history. Within just over 50 years, the technology was sound enough to cross the Atlantic – the first transatlantic paddle steamer got going in 1838. In the same year the paddle steamer began its slow decline, when engineer Francis Smith fitted his underwater wooden propeller to the first propeller-driven ship, the *Archimedes*, so named because the propeller was based on the Archimedes Screw, invented by the Ancient Greek in the third century BC. The Screw was designed to raise water by revolving a screw within a cylinder, and in Smith's propeller the principle was reversed, pushing water back, thus creating momentum.

In 1864, Etienne Lenoir of France came up with the first boat that used an internal combustion engine to power it. Gottlieb Daimler built the first petrol-driven motor boat in 1886. At the time, people were so scared of petrol engines that Daimler added a host of cables to the boat to make it appear as if it was running on electricity!

Car

In the long and complicated history of the development of the motor vehicle, two names stick out: Karl Benz and Gottlieb Daimler. After years of research and development, the first car to be successfully powered by an internal combustion engine was the Benz Motorwagen. Built in 1885 and based on a horse-drawn vehicle, its single-cylinder engine was housed under the driver's seat so it could drive the two rear wheels. In 1886, it was first demonstrated at a speed of 15kph covering a distance of 1km. Two years later, Mr Benz's wife Bertha drove a Benz automobile just over 100km, gaining huge publicity and convincing many that the car was the way forward so far as getting around in style went. Her historic journey is marked by an annual holiday in Germany. Benz's cars were the first to be sold commercially, the first two four-wheeled models being the Victoria and the Vis-à-vis. The first standard model was the Benz Velo of 1894.

In 1889, rival Gottlieb Daimler produced his *Stahlradwagon* (Steel-Wheeled Carriage), the first car with a two-cylinder engine and an impressive top speed of 17.5kph. He died the following year, but his company lived on, and in 1926 the two companies united to form Daimler-Benz, although its two founding members never met.

The cars that truly set the industry on course to produce the sorts of vehicles we drive around today were those produced by Emile Levassor and Armand Peugot from 1891 onwards. They used Daimler engines, crucially placed at the front of the car so that the rear wheels were driven through a clutch and a gearbox.

Catseyes

Driving at night in the early 20th century was far harder than it is today. Headlights were less powerful, and there were no such things as reflective bollards or studs in the road to guide a driver. The only real aid to driving in the dark came from tram tracks, the metal of which would shine in the beams of headlights. But even these were in decline, as trams were being rapidly phased out and the tracks removed by the 1930s. One 1933 evening, Percy Shaw was driving home when, depending on which story you believe, he saw the reflection either of reflectors on a poster by the road, or of a cat's eyes, thus realising he was about to fly from the road's edge. This experience provided him with a flash of inspiration that led to the creation of his road stud. Shaw stated that he tried over a thousand ways of housing reflective studs in the road and, through a lengthy process of trial and error, eventually arrived at a beautifully simple design. The reflectors were mounted in rubber and crucially, when a car ran over a catseye, lenses sunk into the rubber which rubbed over the eye (rather like an eyelid) and cleaned it before it popped back up again. The invention was enthusiastically embraced by the government before the war, just in time for the blackout. Percy Shaw grew infinitely richer, and driving at night grew infinitely safer.

GPS (Global Positioning System)

GPS has many uses, but it became familiar to most of us through in-car use, with a voice telling us to 'turn left' into a road that turns out to be going in the wrong direction up a one-way street! It was developed and used by the US Air Force in 1978, and relied on the first two NavStar satellites. Modern receivers now use four of the 24 satellites flying around space, each satellite having an atomic clock. Received times are compared with actual time and this information, when combined with the current position of the satellites, enables the receiver to accurately ascertain the position on earth. It can then, for instance, tell a car driver when to flick an indicator to exit another roundabout, saving them the bother of simply reading road signs ...

Helicopter

The principle of flight by vertical take-off is often said to have been first conceived by Leonardo da Vinci in 1483. Leonardo certainly laid down a design for such a machine, and it is very helicopter-esque. However, a little-known fact is that 2,000

257

years previously, in Ancient China, there already existed toys with feather propellers that spun like rotating helicopter blades.

Manned helicopter flight, though, was a long way off. The first helicopters had minimal lifting power and tended to fall on to their sides due to a lack of understanding of the principles of the rotor. The first flight of Gyroplane No 1 by two French brothers, Jacques and Louis Breguet, in 1907 took their machine to the dizzy height of 3in above the ground. Another Frenchman, Paul Cornu, managed to get to 30cm later that year, and stayed up for nearly half a minute, but later gave up the dream of helicopter flight. It was the invention of the autogyro by Spaniard Juan de la Cierva in 1923 that really moved things on and anticipated the modern helicopter. 1924 saw the first 'proper' helicopter flight take place. Lasting nearly eight minutes, the machine was manned by Frenchman Etienne Oehmichen over a 1km distance. But the man who takes the credit for the first modern helicopter is Igor Sikorsky, a Russian who developed the VS300 and flew it in America in 1939, just in time for the technology to be used during the Second World War.

Motorbike

Moving on from the bicycle powered simply by human exertion, in 1869 a machine was introduced that ran on steam, an innovation of Frenchman Michaux-Perraux. A four-stroke engine was placed underneath the seat of an ordinary two-wheeled bicycle by Willhelm Maybach and Gottlieb Daimler in 1885 in Germany while they were running tests for their prototype motor-car engine (see page 254). This powerful little two-wheeler caught on. Around the same time in Britain, Edward Butler gave power to the back wheel of a three-wheeled bike by putting an engine on the back of it. In 1894 Hildebrand & Wolfmüller took over where Maybach and Daimler had left off (due to their primary interest having been cars) and fitted a twin cylinder to a two-wheeler. They improved on earlier designs by fitting the engine much lower in the frame, which increased stability, and began selling this first model in 1897. After that, many different companies went into competition, making all sorts of improvements to the original design. But over time a small handful of large companies cornered the market, and they still dominate the motorcycle industry to this day.

The motor scooter came about in the 1940s.

Earlier scooter designs were based on a child's scooter toy – it had no seat and only a board to stand on. Corradino d'Ascanio took the toy's design further and invented the iconic Vespa ('wasp' in Italian) for getting about in the massive airfield where he worked. Originally with a steady board to stand on and a two-stroke engine tucked away at the rear, a seat was soon added and a design classic was born. Originally inexpensive to buy because they were so cheap to make, Vespas soon became desirable and fashionable, and prices changed accordingly!

Parachute

In 1483, Leonardo da Vinci sketched a parachute of sorts. It had a pyramid-shaped canopy, but nothing materialised from what was a rather abstract idea at the time. However, it is André-Jacques Garnerin who made the first parachute and showed great courage in testing it out in 1797. Garnerin's parachute consisted of a container for him to sit in, which was attached by a metal rod to a large piece of canvas hanging overhead. The 2,800-metre descent was a success in that his design had worked and he survived, but he had endured an extremely bumpy ride. When he arrived at his destination – the ground – he was vomiting violently. Thinking that his rough journey may have been due to too much air resistance from the airtight canopy, Garnerin took the small but very effective step of making a small slit in the top of it. Next time around, Garnerin arrived in a more dignified state, and no doubt gave himself a pat on the back for inventing the first effective parachute.

Parking Meter

The despised parking meter was the bright idea of Oklahoma newspaper editor and businessman Carlton Magee. In 1933, Carl found himself on the city's Businessmen Traffic Committee, set up by the Chamber of Commerce to find a solution to the problem of city workers filling the streets with their cars. The congestion was so bad that shoppers had nowhere to park, much to the irritation of shopkeepers. He motioned the idea of using 'Park-O-Meters' produced by the Dual Parking Meter Company, so named because the meters were designed with the dual function of regulating parking and generating revenue. The traffic authority thought it a fine idea and ordered 150 of the devices, which look much like those we use today. Oh, and the Dual Parking Meter Company was owned by Carlton Magee, so it all worked out quite nicely. Needless to say, it wasn't long before city authorities around the world learned how to infuriate their drivers. They first came to Britain in 1958, when 625 were set up in Mayfair, London. Aside from controlling parking and gaining revenue for the government, another 'bonus' that came out of parking-meter installation was the creation of a new vocation – the traffic warden! The first 'meter maids' patrolled the streets of New York and London in 1960.

Railway

Next time you're stuck on an over-packed, overpriced, delayed train, consider the fact that railway wasn't always for tired commuters. It may come as a surprise that the use of railway lines goes back to the Ancient Greeks. In order to achieve audience-pleasing effects in classical theatre productions, rails were employed to bring scenery on to and from the stage at startling speeds. Examples remain at theatres in Sparta and Megalopolis. Such effects would have been impossible to achieve by other means, for transport by rail hugely reduces the power needed to move a load in comparison to wheels on a flat surface. But it wasn't just the luvvies of the ancient world who realised the usefulness of rail – from around 500 BC a railway system was also installed to carry boats over the river Isthmus in Corinth. It remained in use for over a thousand years.

The modern railway finds it origin in mining. From around the 15th century, miners used wooden rails along which they pushed wagons, and the first known example was found in Germany, dating to around 1430. In the English town of Nottingham, mining entrepreneur Huntington Beaumont created a railway line to transport coal from mine to city in 1604, and in 1608 he did the same in Northumberland. Both schemes worked well, but were unprofitable. Nevertheless, he inspired many an imitator and can be said to be responsible for catalysing the development of railway in Britain.

Not much changed for a century or two. Iron rails were introduced in 1767, but it was the combined drive of two men that revolutionised the world of railway. Englishman Richard Trevithick was responsible for the first steam locomotive in 1804, and it ran along the Pen-y-Darren railway in Merthyr, Wales. The tragedy of his life was that people simply didn't believe his ideas were worth anything. Disheartened, poor Richard left the country and eventually died penniless in 1833.

Not that Trevithick's ideas weren't taken up by others. Locomotives were developed in several stages, but it was George Stephenson who stood out from a host of would-be pioneers, opening the Stockton and Darlington Railway in 1825. It was the first railway in the world to carry stock and freight, and it was Stephenson's idea of a national network that marks him out as the 'genius' behind the modern railway. During the following 25 years, railways sprung up all over the country, and by the 1850s over 6,000 miles of railway lines had been laid across Britain, with America, India, Russia and the rest of the world following suit.

River Tunnel

The first tunnel to be dug under a river was the Rotherhithe to Wapping tunnel under the Thames in London. Construction was made possible thanks to French engineer Marc Brunel's tunnelling shield, which supported the tunnel and prevented water from pouring in. Despite this there were still a few accidents along the way – Brunel's son was so severely injured that work ceased for a number of years. Due to this and other setbacks, the tunnel was nearly 20 years in the making, before finally opening in 1843.

Road

Early road surfaces were vastly different to those we drive along today. Nevertheless they served a similar purpose and could be very long. One of the earliest known roads is the Persian Royal Road. Completed around 3500 BC, it covered nearly 3,000km between the Persian Gulf and the Aegean Sea. The Incas also constructed extensive networks of roads, as did the Egyptians who needed to transport huge quantities of materials for the construction of the Pyramids. But most roads were merely big dirt tracks that became impassable in bad weather, and the first road designed to function in all conditions was constructed in Crete. Built by the Minoans around 2000 BC, it had stone paving and was slightly higher in the middle to ensure good drainage. Interestingly, there was a pavement of sorts for pedestrians, but it was in the middle of the road. The largest early road was the Silk Road in China, completed just after 1000 BC. It remained the longest road in existence for 2,000 years.

But from 300 BC it was the Romans who really mastered efficient road building, famously constructing their highways in straight lines wherever possible. The first significant Roman road was the Appian Way (*via appia*). Building began in

312 BC under the reign of Appius Claudius Claecus. It stretched from Rome to Naples in the west, all the way to Brindisi in the east. As the empire grew, so did the Roman road map – supreme roads were laid across Europe and North Africa and, by the first century, the Roman Empire was linked together by some 85,000km of paved roads.

Despite the extensive network of roads in Britain, their quality was variable to say the least. Many were merely dirt tracks that were useless in the rain, and even the paved ones could be a real headache to negotiate, as damaged and loose stones would ruin cart-wheels, and perhaps give the odd twisted ankle to a poor horse. By the 18th century, many years of wear and tear of existing roads, combined with poorly constructed new ones, made drastic improvements a matter of urgency. Thomas Telford, a Scotsman, moved things on significantly with the sturdy foundations he laid building his many miles of roads in the early 1900s. Construction was further improved by fellow Scotsman John McAdam. Rather than using expensive foundations that took a long time to build, McAdam reasoned that, as long as decent drainage and level surfaces could be ensured, there was no reason why roads could not be laid straight on to the soil. Good drainage was ensured when he pioneered the use of gravel and broken stone, which even when compacted allowed water to seep through it. His 'macadamised' roads became the standard method of construction for new roads until the 20th century, when tarmac was invented by Englishman Edgar Hooley. He was a county surveyor, and a dab hand at spotting ruts that commonly appeared in macadamised roads. One day in 1901, he came across an unusually 'rutless' road and discovered that some tar from a wagon had been spilled on it and that, to cover it up, slag from a nearby ironworks had been spread over it. This immediately triggered an idea for improving McAdam's surface. Hooley patented the new road-building technique that used McAdam's surface combined with tar and slag, naming it tarmac in homage to macadamised roads.

Taxi

The first horse-drawn carriages for hire began circulating the streets of Paris and London in the 17th century, although their numbers were regulated by order of Royal Proclamations. Called Hackney Carriages – derived from the French *haquenée*, meaning a hired horse for journeys – they were an immediate success. Inspired by French *cabriolets de place*, called 'cabs' in Britain, Joseph Hansom invented the Hansom cab in the early 1830s. Faster and safer than previous cabs, they became ubiquitous on London streets for many decades until the arrival of the motor cab. Equipped with taximeters, which had been invented in Germany by Wilhelm Bruhn in 1891 to record distance and time in order to put an end to disputes over fares, the first petrol-motor 'taxicabs' were named after his invention. The taxicab first hit London streets in 1903, and New York's in 1907. Soon, they were simply known as 'taxis'.

The Wheel

Wheels and axles are perhaps the most significant invention in history, with so many of the things we take for granted depending on wheels and cogs for their operation. It is hard to imagine a time when the wheel was inessential, yet such a time there was, and a number of early civilisations managed to achieve great things without it. However, before 10,000 BC, the fact that a large part of the world was covered in ice and that most of what was left was either swamp, forest or desert meant that, had wheels been around, they would have been pretty useless.

The truth is that nobody is sure exactly when, or how, the first wheel came about. Indeed, some historians posit that the wheel began life not as a means of transport but as an aid to potters seeking to form well-rounded pots. The evidence for this is the existence of pots from Mesopotamia dating to 5000 BC that display the characteristics of 'thrown' pots – pots made on a wheel rather than being coiled. There also exists a 5,500-year-old clay tablet from Iraq that shows a potter's wheel being used.

Once domestic animals started to be used for farming around 7000 BC, the desire to use them to transport loads efficiently would gradually have evolved. The earliest pictorial depiction of a wheel used in this way is Sumerian, dating to around 3500 BC, a time when the manufacture of goods and their subsequent trade was beginning to flourish. Many historians of the period think that around this time heavy loads were often moved using tree trunks as rollers, often with materials stacked in sledges placed on top of the trunks. It is thought that, over time, grooves were worn into the rollers by the sledges, and at some point someone cottoned on to the fact that, when the small grooves were turned, a corresponding yet larger turn resulted on the part of the roller on either side of the sledge that was touching the ground. This was the beginning of the creation of the axle. A further stage in the wheel's evolution saw the excess wood in the middle part of the log being further cut away to leave an axle and two wheels.

But cutting sections of logs into wheels was laborious and inefficient, so people soon began to make them by fashioning circles from planks and attaching them to an axle. Another breakthrough would have been the realisation that sections could be cut out of a solid wheel, thus making it lighter but retaining its strength. From around 1000 BC, wheels with spokes became the norm, and the evolution of the wheel just kept on rolling.

INDEX